D0014798

SOILS OF THE WEST AFRICAN SAVANNA

This and other publications of the
Commonwealth Agricultural Bureaux
can be obtained through any major bookseller
or direct from:

Commonwealth Agricultural Bureaux
Central Sales, Farnham Royal,
Slough SL2 3BN, U.K.

SOILS OF THE WEST AFRICAN SAVANNA

The maintenance and improvement of their fertility

by

M. J. JONES
(*Institute for Agricultural Research, Samaru, Zaria, Nigeria*)

and

A. WILD
(*University of Reading, England*)

Technical Communication No. 55
of the
Commonwealth Bureau of Soils
Harpenden

COMMONWEALTH AGRICULTURAL BUREAUX

First published in 1975
by the
Commonwealth Agricultural Bureaux,
Farnham Royal, Slough SL3 3BN England

ISBN 0 85198 348 0

Printed in Great Britain by The Cambrian News (Aberystwyth) Ltd.

PREFACE

Higher crop yields are needed in many parts of the world, and the countries of the West African savanna are no exception. Populations are increasing, probably at two to three per cent per year, the farmers themselves want higher yields in order to exchange their surplus for manufactured goods, and for some countries the sale of agricultural produce is the only means of obtaining the foreign currency needed for development. In writing this reveiw, our aims are to show that crop yields can be increased substatially and to show how better soil management can contribute to these increases.

Publicity has recently been given to the prolonged drought in the most northerly part of the savanna, the Sahel zone; but this is a comparatively small part of the total area, and most of the savanna is sufficiently well-watered for one and sometimes two crops to be grown annually. And, as we shall show, the yield potential is high.

Yields can be improved by better soil management, including the use of fertilizers and good soil cultivations; by better crop husbandry, especially timeliness of planting and good weed control; and by using the seed of improved crop varieties. These technical improvements cost money and usually bring about changes in the farming system, changes that may have undesirable economic and social effects, as Frankel (1972) found in India. But over much of the savanna there still remains the opportunity for development wholly directed to the benefit of the people living there. However, in our account we concentrate attention on technical improvements to soil management; social and economic implications we leave to others.

Our account is essentially a review of existing data. It was prompted by the need for a summary of the very considerable volume of research into savanna soils and their fertility that has been carried out over the past thirty or forty years. Although similar environmental conditions occur throughout the region, research has tended to have national boundaries with scientists in different countries apparently unaware of each other's work; and even in any one country the results of experiments have rarely been brought together and compared. It is perhaps partly for this reason that the considerable effort put into agricultural research during this time has not yet yielded the benefits it merits.

In the course of preparing this review we have become aware of the dearth of information from outside the few centres where research is at present concentrated. Regrettably too, some experiments have never been written up, and some results are available only in internal memoranda.

We have, however, found much more data than we expected and our review is therefore a long one. Hence, we advise readers who are more concerned with generalities than with the evidence for them to read the summary in Chapter 12. The earlier chapters are intended to give agronomists and soil scientists, in particular, an account of present knowledge of the fertility and management of savanna soils. It is hoped that these chapters will show where gaps in our understanding exist and how present information is related to the general body of scientific knowledge, for research is useful only when we know how far its results can be extrapolated from the circumstances of the experiment.

We wish to thank Ahmadu Bello University, Zaria, Nigeria for providing us with the opportunity to undertake this and other work on savanna soils, to the Director of the Institute for Agricultural Research in the same University, and particularly the Institute librarian, Miss P. M. J. Edwards for her help in obtaining references. We are also grateful to Professor E. W. Russell for constructive criticism of the typescript and to the Royal Society for a grant towards the cost of publication.

We acknowledge permission of Cambridge University Press to reproduce Figs. 7 and 8. Sources of other data are indicated by reference.

CONTENTS

PART III—FERTILITY REQUIREMENTS FOR HIGHER CROP YIELDS

List of Tables

List of Figures

INTRODUCTION

THE REGION

The area we have taken for our survey is the whole of the cultivable part of West Africa north of the forest zone (Fig.1). Its southern boundary is the forest, and it extends northwards towards the Sahara Desert to the limit of millet cultivation, which occurs with a mean annual rainfall of about 250 mm. Between these two extremes there is an area of about $3 \cdot 5 \times 10^6$ km^2 (over 1 million square miles), known as the West African savanna, stretching for about 4,000 km from the Guinea-Senegal coast in the west to Chad and the Central African Republic in the east. This geographical region occupies about 90 per cent of the land surface between the northern limit and the coast, the rest being forest with a coastal strip of mangrove swamp in some areas. North of the savanna it is generally too dry for rain-fed crop production.

As might be expected from its size, the savanna is not uniform. For example, at the forest boundary the annual rainfall may be as much as 1500 mm and the length of growing season, as defined in Chapter 1, about 260 days; at the northern boundary the rainfall is about 250 mm and the growing season only 55 days. These climatic differences affect the nature of the vegetation, the kind of crop that can be grown, and the potential crop yield. The objectives of soil management also vary; whereas in the south provision must be made for the disposal of surplus water, conservation of water is increasingly important towards the north.

However, although the climate is not uniform, the decreases from south to north of annual rainfall and length of growing season are sufficiently regular for generalizations to be made with some confidence. Variations of other factors that affect crop growth, namely temperature and solar radiation are similarly regular, and day length differences are small. Altitude varies little, and only small areas lie above 1,000 m.

Some common features can also be recognised in the soils. Nutrient supplies are usually poor or very poor, and clay and organic matter contents low, so that soils are poorly buffered and have low capacities for holding exchangeable cations. In addition, agricultural development is at a similar stage of development nearly everywhere, mainly subsistence farming but with an expanding cash crop element. Most farms are small, and there are few plantation crops; yields are low and fertilizers little used.

We have excluded the forest zone, where, because of the less severe dry season, perennial crops such as cocoa, oil palm and rubber can be grown. The direction of agricultural development there is likely to be different from that in the savanna, where farming is restricted almost entirely to annual crops, although many of the basic problems, economic, agricultural and even specifically soil, may be similar.

A final point on the selection of our survey area is that we have followed the precedents of Cocheme and Franquin (1967 a, b) and Morgan and Pugh (1969) in choosing a natural region rather than an individual country. We have done this in order to compare research findings from different countries within the savanna and where possible to show that these findings, whatever their country of origin, can be applied in other parts of the region where environmental conditions are similar.

DEFINITION OF TERMS

(a) *Soil fertility* We follow Cooke (1967) and distinguish soil fertility from soil productivity:

"Soil fertility is concerned with the ability of soil to supply enough nutrients and water to allow the crop to make the most of the site. Soil productivity integrates both the climatic potential of the site and the fertility of the soil".

For our purpose we need to expand Cooke's statement on soil fertility to include those physical properties that affect crops directly or indirectly, like aeration, infiltration rate, water holding capacity, mechanical composition, capping, and also soil erosion.

The study of factors of soil fertility, and of nutrient supplies in particular, becomes of practical value when related to crop yields. But yields depend not only on soil fertility and the climatic potential of the site; they also depend on the choice of a suitable crop variety, on timely planting, on correct planting density and on control of weeds, insects, and diseases, which together we describe as crop husbandry. As we shall see later, one difficulty in relating soil fertility to crop yields is that in some experiments there has been a low standard of crop husbandry. Yet for improved yields improved husbandry is essential. This topic is largely outside the scope of our review, but its importance is above dispute. It will be considered, though briefly, in relation to crop response to fertilizers (Chapter 10).

(b) *Soil management* The aim of soil management is to control and improve soil fertility to give optimal crop production. Soil management includes any kind of manipulation of soil properties. It includes cultivations made either in the interests of soil conservation or improved crop yields. It includes the use of fertilizers, manures and other soil amendments such as lime to change soil nutrient status and also, in the case of organic manures, to change soil physical conditions. And it includes the use of irrigation water in drier areas and the control of salinity and alkalinity. The term 'soil management' implies that soil fertility is under our control, and that it can be manipulated advantageously. To do so will become increasingly important as agriculture becomes more intensive.

SEQUENCE OF PRESENTATION

Part I describes the general environmental conditions of the savanna: climate, soils, and the present pattern of farming.

Part II examines first in some detail the existing fertility of the soils, their physical and chemical properties and nutrient status; and secondly, the relationships between these properties and traditional farming practices, such as fallowing, bush burning and intercropping.

Part III reviews recent experiments aimed at improving and maintaining soil nutrient supplies and creating satisfactory physical conditions and concludes with a summary and brief assessment of the prospects for maintaining soil fertility under higher productivity systems.

One theme of our review is the contrast between actual and potential crop yields. This contrast is unfortunately only too apparent in the neighbourhood of research stations; whereas the ordinary farmer gets x units of yield, over the fence on the research farm yields may often be 5x or 10x. Such differences arise from other factors besides soil fertility, but, as our review will show, substantial increases in farmers' yields are unlikely to be achieved without improvements in fertility. Prices for produce will determine the extent to which it is economic to invest money in higher soil fertility, but there are good reasons for thinking that, if necessary, crop yields from the savanna could often be increased at least fivefold.

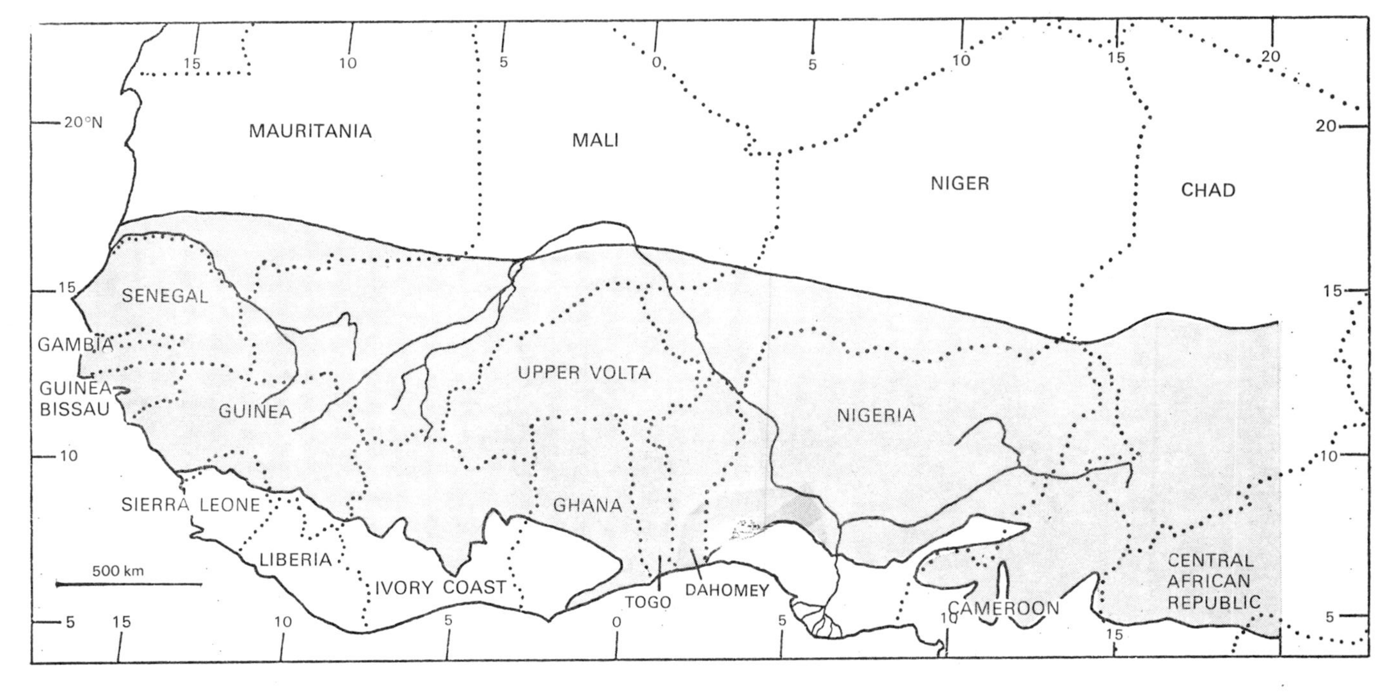

Fig. 1. The area of the West African savanna (shaded). The northern boundary is the limit of cultivation (the 250 mm isohyet), and to the south is the forest boundary (Aubreville *et al.* 1959).

PART I—THE PHYSICAL AND FARMING BACKGROUND

Chapter 1

CLIMATE

1. RAINFALL

(a) *General*

In the West African savanna, according to Kendrew (1961) and Griffiths (1971), there are two dominating air masses, the northerly dry harmattan which comes from the Sahara, and the south westerly monsoon of humid oceanic air. They meet at the Inter-Tropical Convergence Zone which moves north and south once each year. It is furthest south in January, when the harmattan blows almost to the coast, and reaches its northern limit (15—25° N) in August. In the north therefore the dominant wind for most of the year is the northerly harmattan, and in the south it is the south west monsoon. This results in a short wet season in the north and a long one in the south.

The seasonal distribution of the rainfall is given in Table 1 for four stations. In general the annual rainfall and the length of rainy season both decrease from south to north (Figs. 2 and 3). South of latitude 8° or 9° N there are two rainy seasons, the first between April and July and the second, usually with less rain, between September and November. North of this latitude the wettest month is August. The biggest exception to the regular north-south trend is the relatively dry area in south eastern Ghana, southern Togo and Dahomey, where the annual rainfall near the coast drops to 750 mm and savanna vegetation occurs. Another irregularity is that in Senegal the isohyets are close together, and the steep rainfall gradient is associated with a high variability of annual rainfall, particularly in the drier areas (Cocheme and Franquin, 1967a, b; Charreau and Nicou, 1971a).

More detailed information on the annual rainfall distribution and length of rainy season has been given for the northern part of the region by Cocheme and Franquin (1967a, b), for Ghana by Walker (1962), for Senegal by Charreau and Nicou (1971a) and for Nigeria by Walter (1967) and Kowal and Knabe (1972).

Many authors have remarked on the aggressiveness of the rainfall; much of it, particularly in more northerly areas, falls in high intensity storms. Cocheme and Franquin's estimate of 4 mm/h as the mean rainfall intensity

TABLE 1

Monthly amounts of rainfall (P) and potential evapotranspiration (Et) at four savanna sites (Cocheme and Franquin, 1967 a & b)

	Zinder, Niger 13°46′N, 8°58′E			*Kano, Nigeria* 12°00′N, 8°31′E			*Kaduna, Nigeria* 10°28′N, 7°25′E			*Ilorin, Nigeria* 8°32′N, 4°34′E		
	P mm	Et mm	(P-Et)	P mm	Et mm	(P-Et)	P mm	Et mm	(P-Et)	P mm	Et mm	(P-Et)
Jan.	0	138		0	129		0	195		10	117	
Feb.	0	153		0	139		3	186		18	123	
March	0	185		2	176		15	196		64	140	
April	3	192		8	197		64	174		104	143	
May	27	198		71	205		147	157		170	138	32
June	55	185		119	164		170	122	48	193	122	71
July	153	158		209	136	73	226	106	120	142	107	35
Aug.	232	128	104	311	115	196	292	95	197	132	95	37
Sept.	71	146		137	133	4	290	113	177	257	106	151
Oct.	7	167		14	151		86	134		168	115	53
Nov.	0	144		0	135		5	143		28	130	
Dec.	0	130		0	122		0	172		13	126	
Year	548	1924	104	871	1802	273	1298	1793	542	1299	1462	379

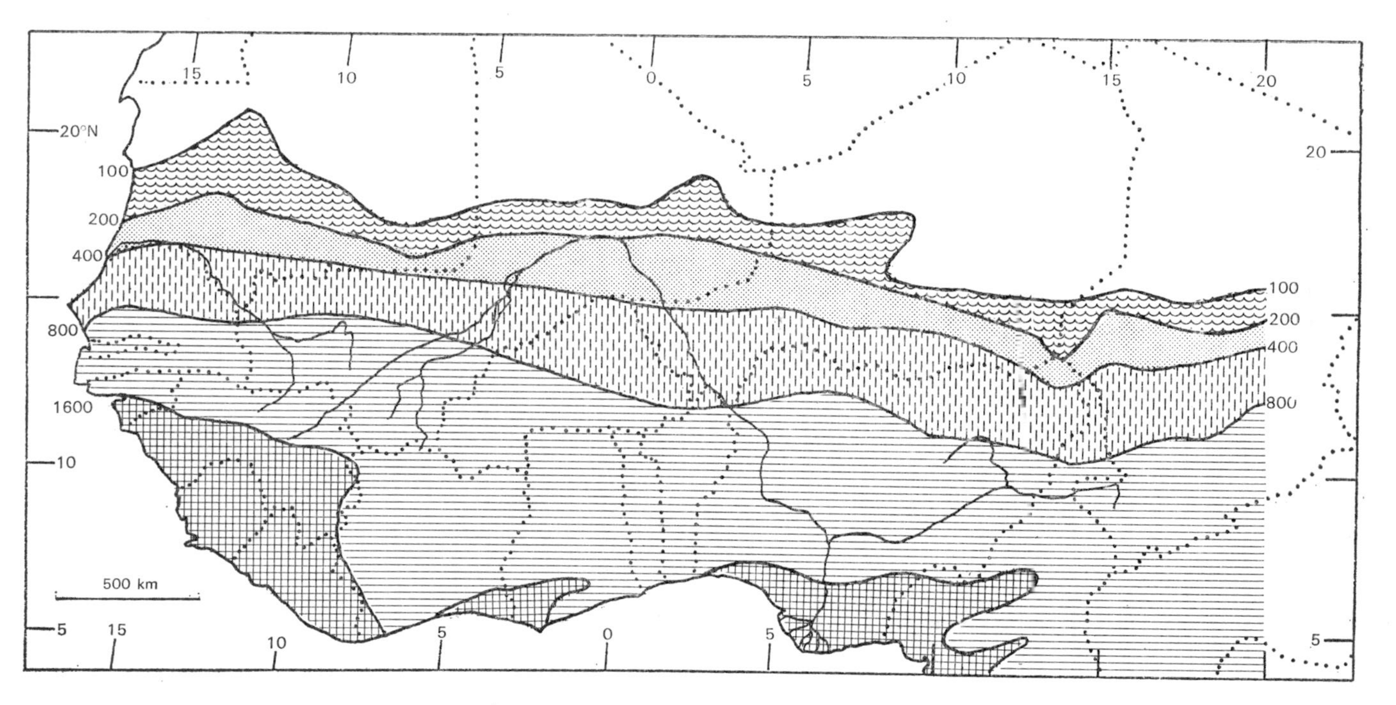

Fig. 2. Isohyets of mean annual rainfall (mm), after Thompson (1965).

in their area of study was questioned by Charreau and Nicou (1971a), who felt that in Senegal, at least, the value is much higher. Pluviograph results showed that half the annual rainfall at Bambey, Senegal fell with an intensity greater than 27 mm/h and a quarter with intensity greater than 52 mm/h; at Sefa, Senegal, corresponding rates were 32 and 62 mm/h, respectively. Peak intensities can be much higher, as shown by data from some individual storms at Bambey:—

12/7/1961: 37 mm in 3 minutes (740 mm/h)
2/9/1961: 33·5 mm in 6 minutes (335 mm/h)
29/7/1961: 18 mm in 3 minutes (360 mm/h)

Although these are figures for exceptionally violent storms, values of 100 mm/h or greater occur in most years at both Bambey and Sefa. Nor are such intensities limited to Senegal. Data, unfortunately, are scarce, but Adu (1972) stated that rainfall intensities exceeding 200 mm/h for short periods are not unusual in Ghana, and in a storm at Samaru, Nigeria, in 1969 about 150 mm fell in two hours, with a peak intensity of 290 mm/h and an intensity range of 120-216 mm/h maintained for 25 minutes (Kowal, 1970d). While storms of such violence are quite rare at any one place, they are undoubtedly an important feature of the climate of the region, at least in so far as it affects the soil; erosion damage at an unprotected site from such a downpour can be catastrophic.

(b) *Rainfall and crop growth*

Annual potential evapotranspiration, ET, is high everywhere and almost always greater than rainfall. In northern Nigeria, assuming a reflectivity coefficient of 0·25, ET ranges from about 1275 mm per annum in the south to 1775 mm in the north and exceeds rainfall north of latitude 7° 30′ N (Kowal and Knabe, 1972). For Senegal, Charreau and Nicou (1971a) quoted the range of 1500-2220 mm per annum, with the ratio of rainfall: ET decreasing from about 0·8 in the south to 0·2 in the north. As Cocheme and Franquin have shown, the pattern is similar across the whole of the northern savanna.

However, although potential evapotranspiration is greater than rainfall on an annual basis, in most years there is almost everywhere a period of water surplus, when rainfall temporarily exceeds ET. Depending on the amount of excess rainfall, profiles can recharge, and drainage and runoff occur.

Cocheme and Franquin (1967a, b) calculated this water surplus, P—ET, for 35 stations, where P is the 30-year mean precipitation and ET the potential evapotranspiration, based on Penman's equation and a surface reflectivity coefficient of 25 per cent. Figures for a number of sites are quoted in Table 1. Both the water surplus, and the period over which it is spread, depend more on P than on ET and therefore decrease with decreasing rainfall from south to north. However, the greater the spread of rainfall through the year, the

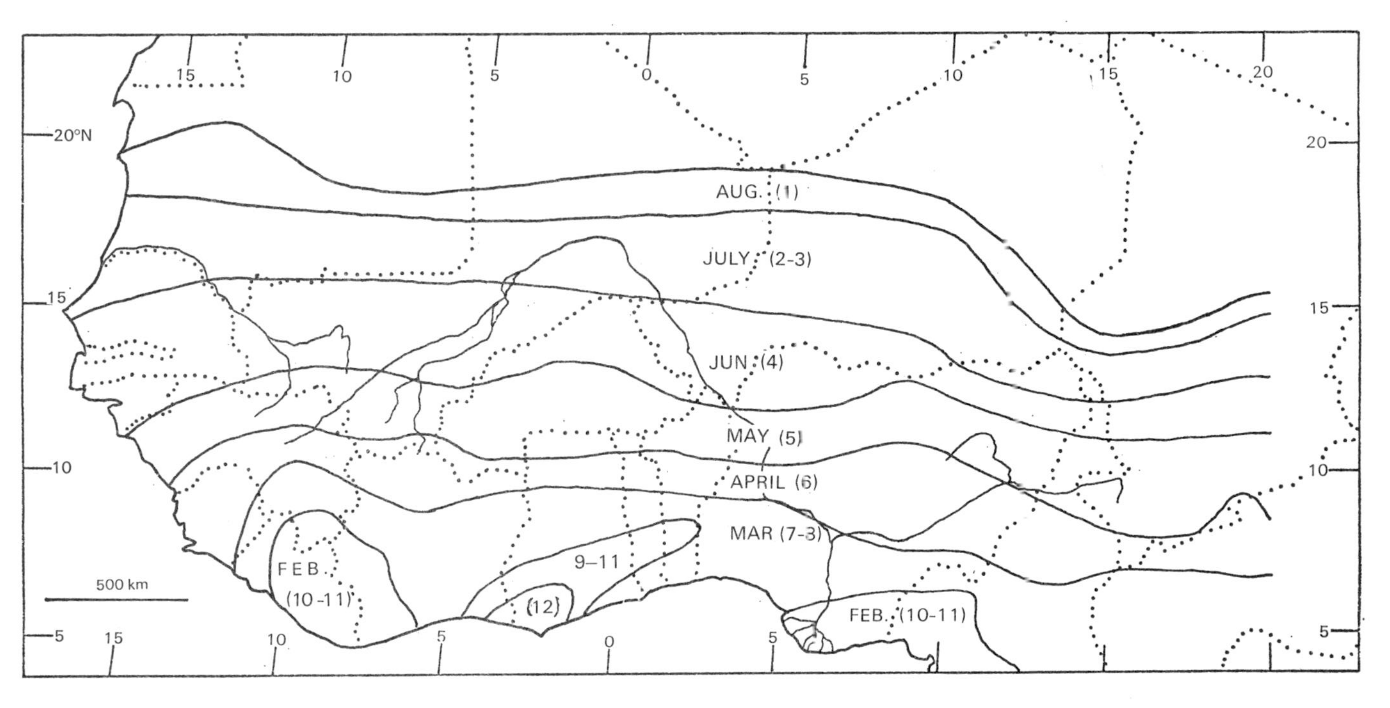

Fig. 3. The first month of the year and the total number of months during which the mean rainfall is 50 mm or more (after Thompson, 1965, with amendments).

TABLE 2

Duration, in days, of the different periods of water availability at four sites in Senegal (Charreau and Nicou, 1971a).

Station	*Rainfall,*	*Periods of water-year*					*Growing season*
	mm	P_R	I_1	H	I_2	R	$I_1 + H + I_2 + R$
St. Louis	346	31	16	18	34	4	72
Matam	535	26	23	49	21	17	110
Thies	694	24	18	74	13	25	130
Tambacounda	941	18	17	109	14	23	163

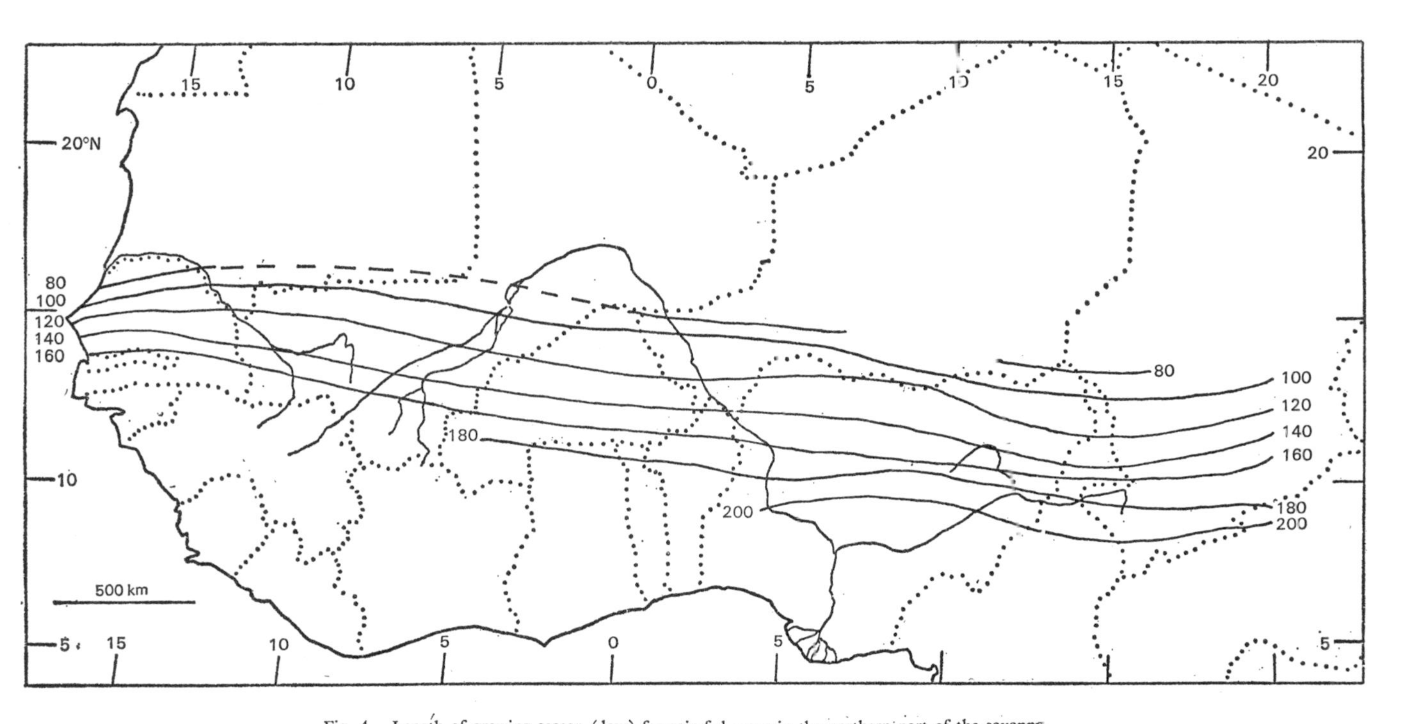

Fig. 4. Length of growing season (days) for rain-fed crops in the northern part of the savanna (Cochéme and Franquin 1967b).
Note: length of growing season is defined as length of first intermediate plus humid periods, plus period of use of stored water (see text).

smaller the excess. For instance, Kaduna (Nigeria) with an annual rainfall of about 1300 mm has a calculated water surplus of 542 mm spread over four months, whereas Ilorin, further south, but with a similar rainfall falling over a longer season, has a surplus of 379 mm spread over six months. Such differences may be expected to affect the magnitude of the erosion risk and leaching rate at different sites. In general, however, the main trend is for the water surplus and the period over which it is spread to decrease towards the northern boundary of the region.

Various authors have found it convenient to divide the "water-year" into well-defined periods according to the probable availability of water to crops (Bonfils et al., 1962; Cocheme and Franquin, 1967a, b; Kowal, 1969; 1970a; Charreau and Nicou, 1971a). Although the systems used vary slightly, the consensus is towards the following:

1. Preparatory period (PR); rainfall, lying between ET/10 and ET/2, is sufficient to permit cultivation but not sowing.
2. First intermediate period (I_1); rainfall, lying between ET/2 and ET, is sufficient for sowing and early growth.
3. Humid period (H); rainfall greater than ET.
4. Second intermediate period (I_2); rainfall between ET/2 and ET.
5. Reserve period (R); rainfall less than ET/2; maturing crops are wholly or partly dependent upon soil moisture reserves.

The lengths of these periods at four sites in Senegal are shown in Table 2. Whereas the length of the humid period and the total length of the growing season (Fig. 4) increase southwards with increasing rainfall, the lengths of the preparatory period and the second intermediate period both decrease. The shortness of the season in the north clearly limits the range of crops that can be grown, but the shortness of the preparatory period in central and southern areas is no less a severe constraint on management. Over the savanna region as a whole the length of this preparatory period ranges from more than 50 days to less than 20 days, being greatest around the 600 mm isohyet in the east (Niamey, Katsina, Ndjamena) and least in southern and central Senegal and Gambia (Cocheme and Franquin, 1967a, b). This fact underlies the research emphasis that has been given to soil preparation in Senegal, in particular to soil cultivation at the end of the season.

Lengths of the periods of the water-year have also been calculated for a large number of sites in northern Nigeria by Kowal and Knabe (1972), although their system varied slightly from that outlined above. The humid period, taken to run from when the soil profile is recharged (after rainfall of 415 mm $\pm$ 75 mm) until rainfall falls below ET, ranged from 0 to 100 days according to latitude (Table 3). All rain falling during this time in excess of evapotranspiration was assumed to percolate through the profile and contribute to deep drainage (that is, runoff was assumed to be negligible).

The amount involved is less than (P-ET) because of the allowance made for the moisture deficit existing in the soil at the beginning of the

TABLE 3

Length of humid period and estimated percolation at twenty stations in northern Nigeria (Kowal and Knabe, 1972).

Station	*Latitude*	*Longitude*	*Mean annual rainfall (mm)*	*Apprx. length of humid period (days)*	**Estimated through percolation (mm)*
Lokoja	7° 48′ N	6° 44′ E	1181	60	191
Makurdi	7° 44′ N	8° 32′ E	1317	90	329
Ilorin	8° 29′ N	4° 35′ E	1275	80	267
Ibi	8° 11′ N	9° 45′ E	1133	40	121
Mokwa	9° 18′ N	5° 04′ E	1069	40	124
Bida	9° 06′ N	6° 01′ E	1227	100	397
Minna	9° 37′ N	6° 12′ E	1329	90	430
Jos	9° 52′ N	8° 54′ E	1410	100	567
Yola	9° 14′ N	12° 28′ E	977	20	79
Yelwa	10° 53′ N	4° 45′ E	1051	60	221
Kaduna	10° 36′ N	7° 27′ E	1286	80	401
Bauchi	10° 17′ N	9° 49′ E	1102	60	297
Samaru	11° 11′ N	7° 38′ E	1114	60	263
Potiskum	11° 43′ N	11° 02′ E	804	30	69
Maiduguri	11° 51′ N	13° 05′ E	704	20	30
Gusau	12° 10′ N	6° 42′ E	1018	60	295
Kano	12° 03′ N	8° 32′ E	884	40	115
Nguru	12° 53′ N	10° 28′ E	552	0	0
Sokoto	13° 01′ N	5° 15′ E	751	0	0
Katsina	13° 01′ N	7° 41′ E	740	30	33

* assumes an initial soil moisture deficit of 200 mm.

season. Figures in Table 3 are based on an assumed deficit of 200 mm, although at any one site the true deficit will vary according to soil type and vegetative cover. However, it is clear that in an average year deep drainage occurs over most of the area surveyed.

Kowal (1969) gave values of actual moisture deficits under various covers at Samaru:

Previous season's cover	*Profile moisture deficit at end of dry season (mm)*
Sorghum	140
Groundnuts	160
Cotton	199
Grass	239
Fire-protected bush	271

Smallest deficits occurred in soils under crops (e.g. groundnuts) harvested early the previous year, and greatest deficits under natural vegetation, especially where this included deep-rooted trees and bushes which continued to transpire through part or all of the dry season. The small moisture deficit under sorghum seems anomalous, though it was supported by other data.

Differences between types of cover in total annual water utilization may have important consequences for the water balance in the profile when the pattern of land utilization is changed. At Sefa, during the fifteen years following the clearing of woodland for arable cropping the water-table rose eight metres. This occurred in spite of high rates of runoff from cultivated land (estimated at 10-50 per cent of rainfall) and almost none from woodland (Charreau and Fauck, 1970). Almost certainly the explanation lies in the far greater dry season evaporation from the natural vegetation. The eight metres rise in water-table represents an annual accumulation equivalent to 160 mm rainfall, although this underestimates the difference in evapotranspiration because it makes no allowance for the disparity in runoff.

This radical modification in water regime is reported to be affecting the soil at Sefa. Pedological observations in 1968 showed appreciable changes in soil profiles from what they had been in 1950. There had been a general accentuation of hydromorphism, tending to modify all soils towards a ferruginous type with mottles and concretions; oxide-bonded aggregates, or "pseudo-sand", had disappeared, and there had been a general increase in compaction and hardness.

The Sefa site is in moist savanna, where the natural flora contains many woody species, some only semi-deciduous. In more arid areas differences in soil moisture regime under crop and bush may be much smaller. Audry (1967) compared evaporation from crops and natural savanna at Dilbini in Chad during two growing seasons, 1964 and 1965 (rainfall 645 and 453 mm, respectively). In neither year did the wetting depth exceed 210 cm under savanna vegetation, but under groundnuts in 1964, 48 mm drained below this depth and 37 mm was left stored in the profile at the end of the season. Under millet in the second, drier year there was no drainage, and 21 mm of the stored water was used.

These two examples illustrate the main problems of water management in savanna agriculture. As a first principle, runoff must be controlled to avoid erosion. In dry areas this presents no additional difficulties: practices must be adopted which encourage maximum infiltration and increase water storage in the soil. However, where rainfall is high, techniques which conserve water increase leaching, with a consequent loss of nutrients, and may also lead to hydromorphism. At the same time, the uncertain nature of the rains in the preparatory and first intermediate periods requires maximum infiltration early in the season, but controlled runoff later. These problems will be discussed in greater detail in Chapter 11.

An important feature of the climate is the occurrence of droughts. Where average rainfall is only just sufficient for a short-season crop, e.g. millet, there will be some below-average years when the crop fails. Similarly, in drier areas further north on the fringes of the desert, a failure of the rains results in starvation of the nomadic cattle herds because there is insufficient grazing. There are not enough long-term data to be dogmatic, but it appears that every few decades there is a run of dry years which, coming after years of reasonably good rainfall, causes a grave food shortage and decimates the cattle herds. Until long-term records are assembled not even tentatively safe limits for stock and cropping can be given, and of course future climatic trends may not follow those in the past. All that can be said is that rainfall variability is known to increase very greatly when annual rainfall drops below 500 mm (Cocheme and Franquin, 1967a, b), indicating a recurrent problem for both farmer and stockman in these marginal areas.

South of this dry, northern area, annual rainfall is generally adequate for the growth of annual crops; but even here yields may be seriously reduced by a late start or an early end of the rains. Even more damaging is a period of drought shortly after germination while profile moisture reserves are still low, for if the newly-planted crop fails it may be too late to replant. From an analysis of the 30-year rainfall records at Zinder (Niger), Cocheme and Franquin (1967b) concluded that the first safe planting date was the start of the first intermediate period I_1 (see above). However, even in areas with higher rainfall than Zinder periods of 7-14 days without rain can occur at least once in the growing season, and these may cause depressions in crop yield.

2. SOLAR RADIATION AND TEMPERATURE

Annual incoming (global) radiation is highest in the north of the savanna at 180-190 kcal/cm^2 and decreases southwards to about 140 kcal/cm^2 (Cocheme and Franquin 1967a, b). The lowest daily values occur in periods of greatest

TABLE 4

Air and soil temperatures (monthly averages, 1954-67) at Samaru, Nigeria, 11° 11′ N., 7° 38′ E, (Inst. Agr. Res., Samaru, 1968)

	J	F	M	A	M	J	J	A	S	O	N	D	Mean
Screen air temp. (°C)													
Max.	30	32	35	35	33	30	28	27	29	31	31	30	31
Min.	14	16	19	22	21	20	19	19	19	18	15	14	18
Soil temp. at 1600 h (°C)													
At 5·1 cm	31	35	39	40	38	33	30	29	31	32	31	31	33
At 30·5 cm	24	26	29	31	30	28	27	26	26	27	26	24	27

cloudiness, that is, generally in August, and also in the winter months because of the lower inclination of the sun (Davies, 1966). It has also been shown by Davies that at Kano in the cropping period May to September daily solar radiation (cal/cm^2/day) is high, averaging 525, with the highest value of 580 in May and the lowest of 437 in August. Further south at Ilorin the May and August values are respectively 440 and 355 cal/cm^2/day.

As with rainfall, temperatures change fairly regularly in a north-south transect. Average values for Samaru, Nigeria are given in Table 4. Diurnal and seasonal variations are greatest in the north. The daily screen maximum may exceed 44°C in April and May, whereas the minimum in December and January may be as low as 4°C for a few nights and occasionally drop to freezing point. When the harmattan wind carries dust, day temperatures can also be relatively low. By contrast, the screen temperatures at the southern limit of the savanna are generally between 24°C and 29°C at all times. The effect of altitude is small because on a regional scale there is little land above 1000 metres.

Until recently little attention had been given to the extent to which radiation and temperature control crop production. There was some evidence that yields of irrigated dry-season tomatoes were increased if soil temperature were raised by mulching (Quinn, unpublished), and it had been observed that cotton grows slowly at this time of year. In general, irrigated crops may be held back by low winter temperatures in the northern savanna. However, recent calculations by Kassam and Kowal (1973) have demonstrated the importance of wet-season temperatures and radiation levels for crop production in Nigeria. They showed that potential net photosynthesis is 20-40 per cent greater in the savanna than in the forest zone, because of the lower radiation intensities and higher night temperatures in the forest zone; and actual production rates of dry matter and economic yield may differ by as much as a factor of two. No comparisons were made within the savanna, but it is reasonable to suppose that differences of only slightly lesser magnitude exist between sites in the north and south of the region.

Growth may be impeded by high soil temperatures, which at the start of the rains can reach 45-50°C a few millimetres below the surface. Although there are no data from the savanna, it seems likely that such temperatures are too high for optimum seed germination and root growth. At Ibadan, Nigeria (in the forest zone) Lal (1973) obtained a significant benefit from protecting young maize plants from high soil temperatures, by mulching, during the hot March-May period.

3. VEGETATION ZONES

The dry season is too long to allow a continuously high rate of water loss by evapotranspiration. Hence, instead of a leaf canopy losing water all the year round as in the forest zone, the vegetation of the savanna consists of separate trees or woody plants with grass which becomes dormant in the dry season (Hopkins, 1965a; Riley and Young, 1966). It is floristically variable since it extends from a boundary with the forest in the south to a boundary in the north with near-desert vegetation. In general, as rainfall and the length of rainy season decrease passing from south to north, trees become smaller and more widely separated. Although the climatic changes are continuous and the vegetation changes are not always sharp, it is useful to recognize some subdivisions.

The five zones of Aubreville et al. (1959) are shown in Fig 5. Passing northwards from the Savanna-Forest Mosaic in the south there are successive east-west belts of relatively moist undifferentiated savanna (Southern Guinea), Northern savanna with *Isoberlinia spp.* (Northern Guinea), relatively dry undifferentiated savanna (Sudan), and wooded steppe (Sahel). The names in parentheses are those commonly used in anglophone territories (Keay, 1959).

The subdivisions inevitably oversimplify the pattern that is found. The zones reflect decreasing water supply from south to north, but water supply also varies locally with soil depth and proximity to rivers. The subdivisions have been criticised because of lack of a sharp floristic break (Adejuwon, 1969). Nevetheless they have proved useful agronomically, and we shall refer to them later. Some of their characteristics are given in Table 5.

For further information on the ecology of the West African savanna the bibliography of Koechlin (1963) and the books by Phillips (1959) and Hopkins (1965a) may be consulted.

Chapter 2

THE SOILS

Soil survey of the savanna is far from complete. Areas that have been surveyed to a varying degree of detail are listed in the bibliographies of Fournier (1963) and Milne (1968), and useful general accounts have been given by Pullan (1969) and Ahn (1970). The most comprehensive description is that of d'Hoore (1964), at a scale of 1:5,000,000, in the Soil Map of Africa and its accompanying memoir.

D'Hoore used mapping units (Classes) similar to those of Aubert (1965). As these have been widely and successfully used in West Africa and are familar to those working there, they are retained in the present work. Nearest equivalents in the USDA classification (7th approximation) have been suggested by d'Hoore (1968). Some comparisons between the two systems have been made by Aubert and Tavernier (1972), and the difficulties of correlating these and other systems used in Africa have been discussed by Sys (1969).

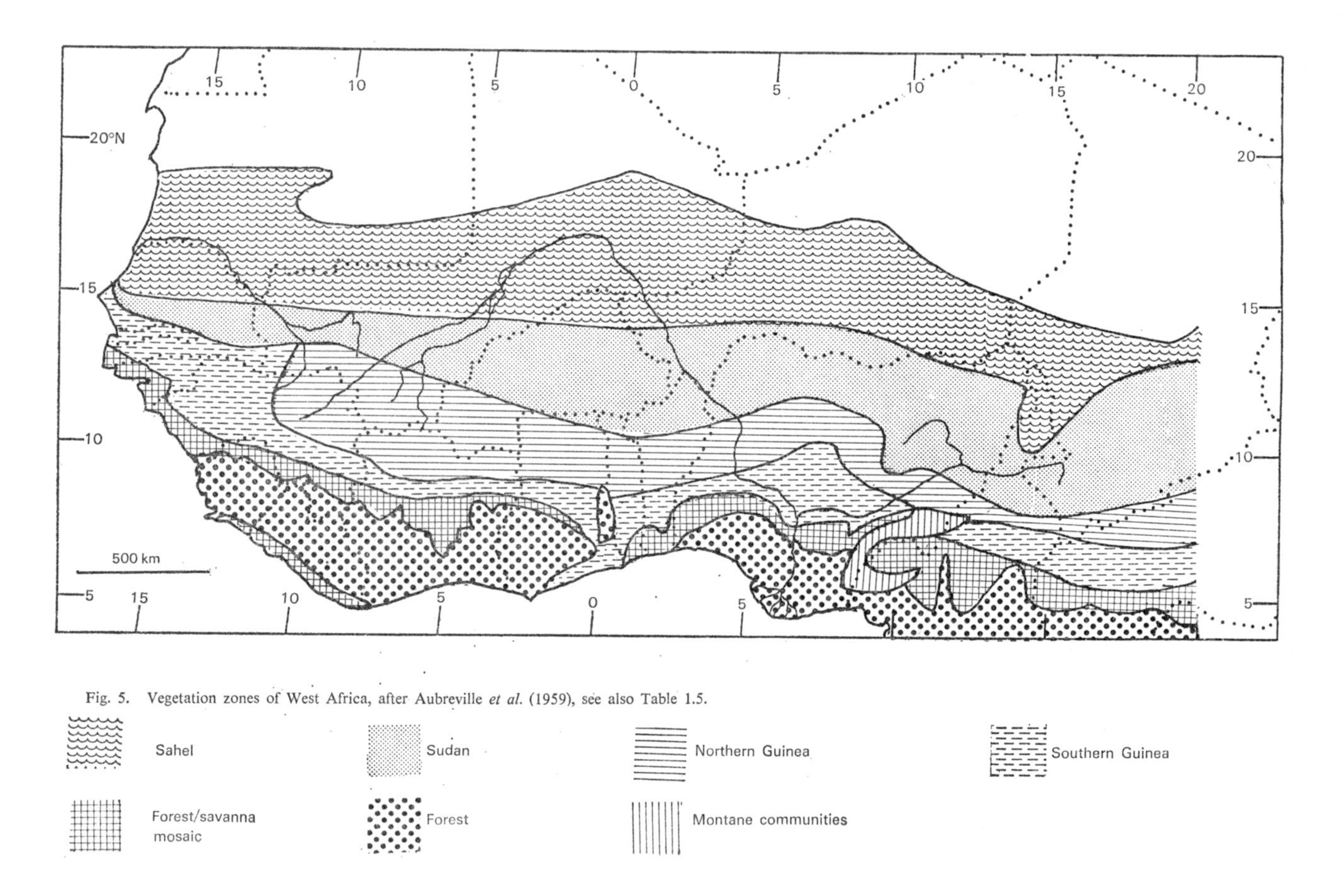

Fig. 5. Vegetation zones of West Africa, after Aubreville *et al.* (1959), see also Table 1.5.

TABLE 5

Ecological zones of the West African savanna (for geographical distribution see Fig. 5).

Zone name:	(1) *Relatively moist* (2) *Southern Guinea*	*Northern Savanna* *Northern Guinea*	*Relatively dry* *Sudan*	*Wooded steppe* *Sahel*
Annual rainfall (mm)	1000—1500	900—1300	500—1000	100—500
Length of rainy season (days)	190— 250	130—190	80— 130	<80
Main soil types	Concretionary Ferruginous; Ferrisols; Ferrallitic.	Leached Ferruginous	Non-leached Ferruginous	Arid brown
Main trees	*Daniellia olivera*	*Isoberlinia spp.*	*Combretum, Acacia, Terminalia spp.*	*Acacia, Commiphora spp.*
Main grasses	*Andropogon tectorum, Imperata cylindrica*	*Hyperrhenia, Andropogon spp.*	*Andropogon gayana*	*Cenchrus spp.*
Main food crops	yams, maize, sorghum	sorghum	millet, sorghum	millet
Main export crops	soya beans, sesame	cotton	groundnuts	—

Note: Zone name (1) after Aubreville *et al.* (1959)
(2) after Keay (1959)

1. GEOLOGY

A number of accounts are available. The following summary is drawn largely from the books of Haughton (1963) and Furon (1963). A bibliography has been given by Dixey (1963).

The rocks of the southern half of the savanna are mainly granites and granitized metamorphic rocks with smaller areas of quartzite, schists, and undifferentiated basement complex. The basement complex rocks show great variation in grain size and in mineral composition, ranging from very coarse-grained pegmatite to fine-grained schist, and from acid quartzite to basic rocks consisting largely of amphibole (Smyth and Montgomery, 1962). A belt of grits and shales runs northwards from central Ghana, and shales and sands occur in southern and eastern Nigeria.

Much of the northern half is covered by unconsolidated or poorly consolidated sands and undifferentiated basement complex, with stretches of alluvium along river valleys. In the west, in Mali, Senegal and Mauritania, there is an extensive area ('Terminal Continental') of marine conglomerates, sandstones, clays and marls, some of which are calcareous.

Volcanic rocks occur on only a relatively small scale, in the Cameroun mountains, on the Jos plateau in Nigeria, and in Senegal. One outstanding feature of the geology of West and North-West Africa is the lack of any extensive occurrence of base-rich primary rocks. Alluvium, colluvium and aeolian deposits are therefore frequently low in plant nutrients.

2 SOIL FORMATION

The land surface of the savanna is old. According to King (1967) the principal surface is Post-African of the late Cainozoic with comparatively small areas of greater age (African), and with younger surfaces close to the major river systems. The two older surfaces have remnants of plinthite (lateritic ironstone) often forming the surface of flat-topped hills. Further geomorphological aspects have been discussed by King (1967), Moss (1968) and Thomas (1969).

Much of the material of present-day soils has been subject to repeated pedogenesis under varying climatic regimes over long periods of geological time, while erosion, colluviation and aeolian deposition have resulted in transport and mixing. Few soils, therefore, are the product solely of the pedogenetic processes seen to be operating today. Nevertheless, certain patterns in soil distribution that relate to identifiable pedogenetic factors, such as climate, parent material, and topography can still be recognised, and these relationships are discussed briefly below.

(a) *Parent material*

From what has been said it will be clear that the influence of parent material has often been obscured by prolonged pedogenesis; indeed, in some cases it may be very difficult to decide what the parent material actually was. Even so, a number of authors have found it possible and useful to relate the nature and fertility of soils to their apparent parent material. Both Leneuf (1954a) and Jenny (1965) divided the soils of Upper Volta according to the geological formation upon which they are found: fertile brown soils, high in clay and mineral reserves, on Pre-Cambrian Birrimian formations (amphibolite shales, dolerites and diorites), fertile, heavy, base-rich soils on alluvium, but sandier soils of only medium or low fertility on granite and sandstones. Further south, in Ghana, the geological boundary appears to determine the position of the forest-savanna boundary (Ahn, 1970). Soils over Birrimian rocks carry forest, but the poorer sandier soils of low water-holding capacity over sandstone support only savanna vegetation.

A relationship between soil texture and parent material is widely recognised: sandy soils over sandstones and granites, with finer textured soils over shales, schists and alluvium. Further, the chemical composition of the rock appears to affect the type of clay formed. In north Cameroun (Martin *et al.*, 1966), in Central African Republic (Boulvert, 1968) and in Ghana (Stephen, 1953) basic rocks were found to favour the formation of 2:1 lattice silicates, including montmorillonite, whereas rocks high in quartz and felspar generally give rise to kaolinite. The cation exchange capacity of soil therefore tends to be higher over basic rocks.

A dependence of soil cation status on parent material may also be demonstrated. For example, in Nigeria, Wild (1971) showed that total potassium averaged 2·18 per cent in savanna soils on basement complex (18 samples) but only 0·08 per cent (4 samples) in those on sandstone; and Boulvert (1968), in Central African Republic, showed that soils over para-amphibolite were much richer in exchangeable and total bases than those over migmatites.

It may be noted here that large differences in clay type and nutrient status exist not only between soils on different groups of parent materials but between soils on similar parent materials (e.g. schists or gneisses) with different mineralogical composition. An abundance of ferromagnesian minerals favours a high soil base status, but it may also encourage profile induration. Pan formation occurs most commonly over basement complex rocks rich in iron and less readily in sandstone soils and over granites poor in iron (Higgins, 1965a, b; Boulvert, 1968).

The transport and mixing of soil materials is most apparent in the northern part of the savanna where, close to the desert, much of the parent material is sand of aeolian origin. This generally gives soils poor in nutrients and of low water-holding capacity. Further south are areas of finer material, often described as loess or drift (Higgins, 1963), deposited over basement complex. An important factor in the fertility of these soils is the depth of the deposit, for this controls the degree of influence of the underlying rock on the biological accumulation of nutrients in the topsoil.

The alluvium of the river valleys is of diverse origin, but as the predominant primary rocks of the region are mostly granites, schists and gneisses, and outcrops of basic rocks are few, most alluvial soils are low in nutrients. One exception to this pattern is the Vertisols of the Chad basin, which have been influenced by the basalt massifs of Cameroun.

(b) *Climate and vegetation*

Superimposed on the background of parent material is the effect of climate. There is a marked zonation of soils across West Africa, following, like the vegetation belts, the trend of the isohyets: Ferrallitic Soils and Ferrisols in the south, Ferruginous Tropical Soils in central areas and the Brown and Reddish Brown Soils of Arid and Semi-Arid Areas in the north (Figure 6).

The basis of the differentiation between these soil types is the intensity of leaching and weathering, largely climatically determined, to which the soil materials have been subjected. A variety of interacting factors are involved: hydrological ones such as rainfall and the volume of through leaching, the period each year the profile is wet, and the rise and fall of the water-table; temperature; and, not least, vegetation. Few of these have been studied in any detail. The clearest account of our present knowledge of them is that of Ahn (1970). Here we mention only a few points that appear most relevant to fertility:—

(i) As one goes northwards across the savanna, profiles become shallower and, depending on parent material, richer in weatherable minerals. This trend has been attributed both to a declining weathering intensity with decreasing mean annual rainfall, and to a greater exposure to erosion as the vegetation becomes sparser. One consequence of shallow profiles is that the deep-rooted components of the savanna vegetation may often be in contact with weathering minerals, so that the soil-plant nutrient cycle is not as closed as it is on the deeply weathered profiles of the forest zone.

(ii) During prolonged and intensive weathering, losses of silica are proportionately greater than those of the sesquioxides of iron and aluminium, and the silica: sesquioxide ratio of the clay fraction can be used as an index of the degree of soil weathering. In some very highly leached soils under forest the clay fraction may be largely sesquioxide. In the savanna, kaolinite predominates in Ferrallitic and Ferruginous Soils, but in the soils of the arid and semi-arid areas 2:1 lattice silicates are usually present. This has implications for soil physical properties and cation exchange capacity.

(iii) Ferruginous Soils dominate the savanna between the 500 and 1200 mm isohyets, but Ferrallitic Soils, though more typical of the forest zone, also occur. However, whereas Ferruginous Soils are characterized by

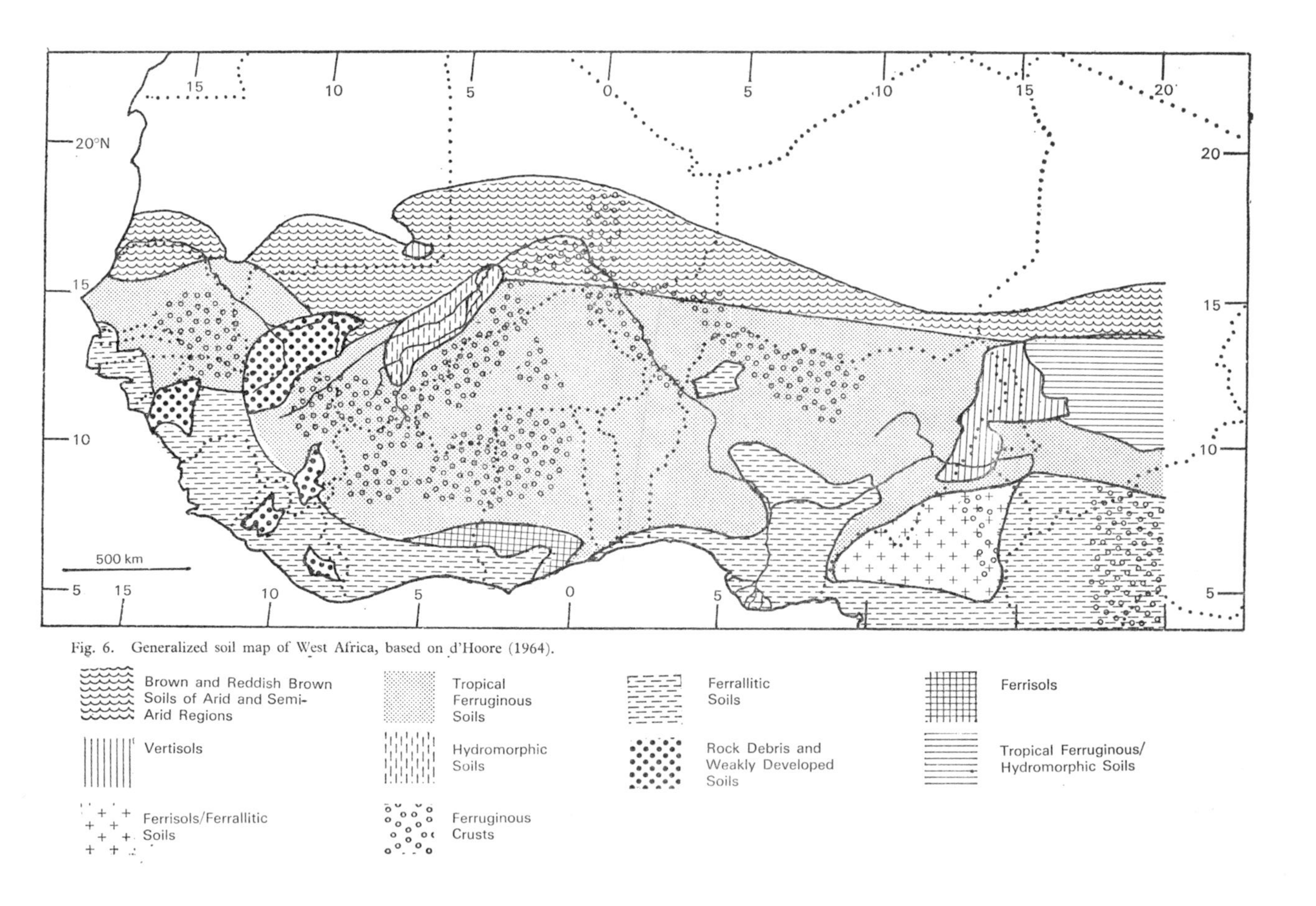

Fig. 6. Generalized soil map of West Africa, based on d'Hoore (1964).

a separation of iron into mottles, concretions or hardpan (plinthite), in Ferrallitic Soils, although some segregation of iron occurs, it rarely forms plinthite. Factors behind the different development of these two soils include, according to circumstances, parent material, previous weathering history and topography (see below), but also the length and intensity of the dry season.

(iv) The big difference in the density of the natural vegetative cover between north and south affects both the extent to which land surfaces are exposed to natural erosion and the degree to which through leaching is reduced by evapotranspiration.

(v) The quantity of nutrients cycled through the vegetation is greater in the forest zone than in the southern savanna (Nye and Greenland, 1960), and it is probably safe to assume that it becomes progressively less as vegetation becomes sparser in more northern drier areas. Even so, it may still be appreciable. For instance, Maignien (1959), in discussing semi-arid soils in Senegal, referred to the 'pumping' of nutrients to the soil surface by plants, and observed that the natural vegetation seemed to create a minimum threshhold level of potassium in the topsoil, despite the poverty of the parent material in that element.

(c) *Topography*

Much of the savanna has a gentle or almost flat relief, although occasional inselbergs or bornhardt domes give prominent landscape features with steeper slopes, as do outcrops of ironstone (plinthite). The effect of relief on the nature of the soil was discussed in general terms by d'Hoore (1964) and later, in more detail, by Moss (1968). In the words of d'Hoore: "all country which has relief consists of zones of departure, of transference, and of accumulation, the limits of which can be peculiar to each transferable constituent or to each group of constituents of comparable mobility. This applies equally to the most simple landscapes and to great geomorphological units". There exist also level zones, covered with altered materials, whose position, usually in the centres of plateaux, allows neither gain nor loss except by wind action. As Moss (1968) pointed out, each zone implies certain characteristics in the soils found there. Moss also discussed the relationship between soil and topography in terms of the catena concept; different slope forms are recognized, and with each is associated a particular type of toposequence.

The subject is a complex one, but in essence there are two basic concepts:—

(i) Wherever there is relief the processes of colluviation are active, stripping soil material from upper slopes and depositing it on pediments and valley bottoms.

(ii) Where geology and climate are uniform or change systematically, repeating patterns of topography give rise to repeating patterns of soil and vegetation.

For illustration we may cite a few examples:—

1. Eutrophic Brown Soils occur only where there is a constant rejuvenation of the parent material, either by the deposition of alluvium or by the loss of surface soil by erosion at a rate comparable to that of the progression of weathering into unweathered rock (d'Hoore, 1964).

2. Vertisols of topographic depressions form in poorly drained areas on parent materials enriched directly or indirectly from the surrounding higher land (d'Hoore, 1964).

The existence of both these soils depends on a transference of soil materials, determined by topography.

3. As already mentioned, one important characteristic of Ferruginous Soils is a tendency towards hardpan formation. In some cases the cause may be a movement of iron into the profile from external sources, either through a seasonal rise in the water-table or through lateral leaching. This process depends on topography (Maignien, 1961). In areas where Ferruginous and Ferrallitic Soils occur side by side the former seem to be favoured by slightly hydromorphic conditions (Chatelin, 1969); and toposequences have been described in which Ferrallitic Soils on well-drained upper slopes gradually shade into Ferruginous and finally Hydromorphic Soils on lower slopes and valley bottoms (Boulvert, 1968; Leveque, 1969).

3. DISTRIBUTION OF SOIL CLASSES

On the map of d'Hoore (1964) the soils of the West African savanna are divided into about ten main Classes. An estimate of their relative distribution, based upon the frequency of their occurrence along the latitude lines (at 1° intervals) of his map, is given in Table 6. For this purpose the savanna was assumed to cover the area south of a line joining the mouth of the Senegal river with the northern shore of Lake Chad (approximately the 400 mm isohyet), north of the forest/savanna boundary, and west of the Nigeria/Cameroun border.

Each mapping unit indicates only the predominant soils within it. Some soils therefore, which occur in small widely scattered areas, may be underestimated, for example the Hydromorphic Soils of innumerable small valleys. In addition, many of the mapping units represent associations of two or more soil Classes and, following d'Hoore, it was assumed that each Class was equally represented within the unit in order to enumerate the frequency of occurrence of each one. Some associations, especially those involving weakly developed soils on ferruginous crusts, occur so frequently that they have been separately recorded.

It will be seen that the dominant soil of the savanna is the Ferruginous Tropical Soil which, including associated areas of ferruginous crusts and soils formed over them, covers about 58 per cent of the region. This compares with 11 per cent for Africa as a whole (d'Hoore, 1964). These soils form a

TABLE 6

The relative distribution of soil types in the West African savanna based on the Soil Map of Africa (d'Hoore, 1964).

		Approximate percentage of the total area:— *Sub-totals*	*Total*
1.	Rock, rock debris and weakly developed soils, including ferruginous crusts.	—	9·2
2.	Vertisols and similar soils.	1·3	
	Vertisols and similar soils associated with weakly developed soils on ferruginous crusts.	0·6	1·9
3.	Brown and Reddish-Brown Soils of Arid and Semi-Arid Areas.	3·6	
	Brown and Reddish-Brown Soils of Arid Areas associated with ferruginous crusts.	1·0	4·6
4.	Eutrophic Brown Soils.	1·0	
	Eutrophic Brown Soils associated with weakly developed soils on basic rocks.	1·7	2·7
5.	Ferruginous Tropical Soils on sandy parent material.	7·2	
	Ferruginous Tropical Soils on crystalline acid rocks.	26·6	
	Ferruginous Tropical Soils on undifferentiated parent material.	2·3	
	Ferruginous Tropical Soils associated with ferruginous crusts and weakly developed soils on ferruginous crusts and other parent materials.	22·2	58·3
6.	Ferrisols.	4·1	
	Ferrisols associated with weakly developed soils on basic rocks.	0·2	
	Ferrisols associated with weakly developed soils on ferruginous crusts.	3·0	7·3
7.	Ferrallitic Soils.	7·6	
	Ferrallitic Soils associated with ferruginous crusts and rock debris.	2·5	10·1
8.	Hydromorphic Soils.	—	5·1
9.	Halomorphic Soils.	—	0·6
10.	Calcimorphic Soils.	—	0·1

broad belt running east-west across the savanna. They are bordered on the north by Brown and Reddish-Brown Arid and Semi-Arid Soils, and in the south by Ferrisols and Ferrallitic soils (Figure 6).

Eastwards of the Nigeria/Cameroun border, outside the area for which the figures in Table 6 apply, the savanna extends into Cameroun, Chad and the Central African Republic. Here Ferruginous Tropical Soils are rather less extensive and Vertisols and Ferrallitic Soils are rather more extensive than they are in the western savanna.

4 THE SOIL CLASSES

Topsoil analytical data for the six most important soil Classes, collected from twenty published sources, are summarized in Table 7. Each Class is represented by sites in each of several widely separated areas within the region, so that the data may be taken to give at least a general indication of the level of the chosen soil properties within each Class.

Mean clay contents are low except in Vertisols, Eutrophic Brown and Hydromorphic Soils, and this pattern is repeated for organic carbon, total phosphorus and total exchangeable bases. This relationship is not accidental: soil carbon content is known to be correlated with soil clay content, soil phosphorus occurs mainly in the organic and clay fractions, and these fractions provide also the exchange sites for the exchangeable bases.

However, within each Class there are wide variations. This is particularly so with the Hydromorphic Soils which occur on alluvium of widely diverse origin, but variation is to be expected in all Classes, because the classification is based on the mode and intensity of soil evolution (d'Hoore, 1964). For this reason, in the following brief Class descriptions only broad generalizations can be made about soil fertility.

(a) *Ferruginous Tropical Soils*

These soils occur widely between the 500 and 1200 mm isohyets in areas having a contrasting wet and dry season regime. They are usually formed on parent materials rich in quartz, that is, on the crystalline rocks of the basement complex, on granite and on aeolian and many sedimentary deposits, but not on volcanic, ultrabasic or calcareous rocks (d'Hoore, 1964). Analytical data for three profiles, formed under widely different rainfall regimes, are given in Table 8.

A common feature of these soils is the downward movement of clay within the profile, a process which tends to produce a sandy surface soil, rather low in organic matter and base exchange capacity, and a compact subsoil where the clay has accumulated.

A second feature is the separation of free iron oxides which are deposited in the profile in the form of mottles, concretions, or a ferruginous hardpan. The formation of this relatively impermeable hardpan and the consequent

TABLE 7

Comparison of main soil Classes (samples generally from 0—15 cm depth).

Soil Class	*Soil Order (USDA, 1960)*	*Clay*		*pH (Water)*		*Org. C*		*Total P*		*Total exch. bases**	
		No. of samples	*Mean % and range*	*No. of samples*	*Mean and range*	*No. of samples*	*Mean % and range*	*No. of samples*	*Mean ppm and range*	*No. of samples*	*Mean (me/100g) and range*
VERTISOLS	VERTISOLS	20	51·8 22·0—78·7	30	7·1 4·6—8·8	29	1·00 0·35—3·47	30	194 75—591	29	26·0 8·3—57·6
BROWN AND RED-BROWN SUB-ARID	INCEPTISOLS	9	6·9 3·8—20·6	15	6·8 5·9—7·2	15	0·25 0·07—0·44	15	92 40—194	15	4·3 1·1— 9·3
EUTROPHIC BROWN	INCEPTISOLS	8	19·5 7·9—36·0	21	6·8 5·4—7·2	21	1·00 0·47—3·80	20	161 40—630	21	10·0 3·5—18·5
FERRUGINOUS TROPICAL	Mostly ALFISOLS	195	9·2 0·0—34·0	245	6·2 4·8—8·2	245	0·62 0·09—2·84	181	125 13—560	245	3·1 0·7—11·8
FERRALLITIC	Mostly OXISOLS	33	9·5 0·4—28·8	33	6·0 4·8—7·3	32	0·83 0·21—1·79	22	127 44—314	32	2·3 0·4—10·5
HYDROMOR-PHIC	—	66	24·1 1·5—71·5	86	6·0 4·1—8·4	90	1·10 0·08—4·92	66	182 46—580	89	7·0 1·4—24·5

Sources: Baker *et al* (1965), Bocquier and Claisse (1963), Boulet (1968), Brammer (1962), Brammer (1967), Carroll and Klinkenberg (1972), FAO (1967), Higgins and Okunola (1965), Klinkenberg (1965), Klinkenberg (unpublished), Klinkenberg and Hildebrand (1965), Leneuf (1954a), Magnien (1961), Obihara *et al* (1964), Pullan (1961, 1962a, b, 1970), Quantin (1965), USAID (1962).

*Exchangeable Ca, Mg, K and Na.

TABLE 8

Ferruginous Tropical Soil profiles

		Profile 1 (*Fauck*, 1963)				*Profile* 2 (*Boulet*, 1968)			*Profile* 3 (*Leneuf*, 1954*a*)	
Parent material		*Sands and sandstones of 'Continental Terminal'*				*Aeolian sand*			*Granite*	
Annual rainfall (*mm*)		1350				440			830	
Site		*Sefa, Senegal*				*Oursi, U.Volta*			*Koupela, U.Volta*	
Sample depth, cm		0—6	6—13	13—31	31—79	0—10	20—30	60—70	0—10	50
Particle size distribution, %	2·0—0·2 mm	26·1	31·1	23·3	21·3	32·4	32·8	32·0	30·5	28
	0·2—0·02 mm	58·2	53·7	48·8	36·5	62·1	56·9	51·3	40·0	22
	0·02—0·002 mm	4·0	4·7	4·9	5·3	1·3	2·0	2·3	6·1	9
	<0·002 mm	9·4	9·4	21·5	35·1	4·0	8·1	14·2	21·5	37
pH (water)		6·5	5·6	5·6	6·2	5·9	5·1	5·5	6·4	6·0
Org. C (%)		1·05	0·68	0·38	0·33	0·14	0·11	0·09	—	—
Total N (%)		0·08	0·05	0·03	0·04	0·02	0·02	0·01	—	—
Total P (ppm)		98	53	76	102	13	40	—	184	114
Exchangeable cations: (me/100 g soil)	Ca	2·25	1·70	1·70	2·10	0·83	1·97	2·03	6·32	9·34
	Mg	1·05	0·65	1·05	1·50	0·45	0·08	1·41	2·23	2·58
	K	0·15	0·10	0·05	0·05	0·10	0·06	0·05	0·32	0·33
	Na	0·05	0·05	tr	tr	0·00	0·00	0·01	0·02	0·08
	Sum	3·50	2·50	2·80	3·65	1·40	2·10	3·50	8·88	12·33
C.E.C. (me/100g)		4·80	4·05	4·45	5·15	1·45	2·15	3·70	—	—

seasonal occurrence of hydromorphic conditions in the overlying soil, as in the groundwater laterites of Ghana (Brammer, 1962), may be regarded as the final stage in the development of a Ferruginous Soil. However, the attainment of this degree of development depends less upon the age of the soil and the vigour of the weathering processes than upon the nature of the parent material. Thus there is usually little accumulation of iron or clay in the profiles of soils developed over sandy materials of aeolian or sedimentary origin ('sols ferrugineux tropicaux non ou peu lessives' of the French classification), but over basement complex rocks relatively rich in ferromagnesian minerals ferruginisation is usually pronounced. The influence of external sources of iron on hardpan formation in certain topographical situations was mentioned above.

Ferruginous Tropical Soils are usually fairly shallow soils, and most profiles are less than 150 cm deep. This is of importance over crystalline rocks, where deep-rooting trees can exploit nutrients released by the decomposition of minerals in the parent material. Primary minerals undergoing decomposition may also occur higher in the profile, depending on the parent material, but they are rare in soils formed on sedimentary and aeolian materials. The influence of parent material on the chemical status of these soils can be quite a strong one, as in the profiles in Table 8.

The clay fraction is predominantly kaolinitic, sometimes with small amounts of illite, but montmorillonite is absent. In the surface horizon clay contents are low but very variable (0—25%), again depending on parent material. Cation exchange capacities, therefore, are also low (1—10 me/100 g soil) and depend closely on the soil organic matter status. This in turn is related to rainfall, soil clay content, and previous land use, but rarely exceeds 2—3 per cent and, in the more arid areas, may be less than 0·5 per cent of the soil.

The agricultural value of these soils is usually rated poor to average. In the better soils over basement complex rocks, nutrient levels, except that of phosphate, are at least moderate, but physical properties may hamper efficient agricultural utilisation. Structural stability is generally poor, and the consequent surface compaction and crusting enhance runoff (Maignien, 1961; Kowal, 1968a). Thus the erosion hazard is high (Bocquier and Claisse, 1963; Lamouroux, 1969). The consequences of erosion are particularly grave on soils which have hardpan in the profile, for a continual reduction in the potential rooting depth eventually puts them out of production. Elsewhere, hardpan horizons may induce temporary hydromorphic conditions within the rooting zone during the growing season.

(b) *Ferrallitic Soils*

These soils underlie much of the forest zone of West Africa, but there are extensive outliers in the savanna, notably in Nigeria, where their occurrence under 760 mm mean annual rainfall and a seven month dry season appears anomalous (Tomlinson, 1965). Their presence seems to follow from the nature of the parent material, an acid permeable sandstone deficient in weatherable minerals, although a more humid climate in the past may also have played a part.

Ferrallitic Soils are deeper soils than Ferruginous Soils, mellower, better structured and more porous. Drainage is good and clay content increases only gradually with depth. Horizons are only slightly differentiated with diffuse or gradual transitions, and there is no accumulation of iron; rather, the iron oxides are evenly distributed on kaolinitic surfaces through the profile (Maignien, 1961). Little or no weatherable minerals remain, and the clay fraction consists of kaolinite and iron and aluminium oxides. Analyses of two profiles are given in Table 9.

Leaching is more intense in Ferrallitic Soils than in Ferruginous Soils, and their subsoils tend to show low base saturation; but higher organic matter levels in the topsoil provide more base exchange sites, and these are often moderately well saturated. Nutrient availability, except of phosphate, may therefore be adequate. However, it is important to appreciate that the maintenance of even a moderate level of nutrient availability is dependent on the maintenance of topsoil organic matter and the return of nutrients in the litter.

Ferrallitic Soils are typically forest soils and need both the buffering effect of a tree canopy on the moisture, temperature and physical condition of their surface horizon and the nutrient 'pumping' effect of deep-rooted species to recover nutrients that have been leached (Maignien, 1961). These soils, therefore, are much better suited to forestry and tree crops than to annual crops (Fauck, 1964; Quantin, 1965; Lamouroux, 1969). Nevertheless they are increasingly used for annual crops, and it is unlikely that this trend will easily be reversed.

The Ferrallitic Soils within the savanna region are much less severely leached than those occurring under a higher rainfall. The influence of parent material is still appreciable, and actual fertility varies widely according to parent material (Fauck, 1964; Quantin, 1965). One may compare the two examples in Table 9. Lamouroux (1969) rated Ferrallitic Soils as the best in Togo: "deep, with very good physical properties, rich in nutrients on basic rocks, rather poor on acid rocks, but this poverty is compensated by the large soil volume available to root exploitation". On the other hand, d'Hoore (1964) described the more sandy Ferrallitic Soils developed on loose sandy sediments as among the poorest soils in Africa. A number of profiles of this sort in the derived savanna of eastern Nigeria, described by Obihara *et al.* (1964), had less than 1 me/100 g total exchangeable bases in the topsoil.

As has been said, the physical properties of these soils are good. However, they often occur on moderately steep slopes and, following the removal of the vegetative cover that has previously held them stable, they are as prone to erosion as the Ferruginous Soils.

TABLE 9

Ferrallitic Soil profiles

		Profile 1 (*Quantin*, 1965)			*Profile* 2 (*Quantin*, 1965)	
Parent material		*Quartz sand*			*Gneiss*	
Annual rainfall (mm)		1400			1530	
Site		*Mouka, Central African Republic*			*Grimari, Central African Republic*	
Sample depth(cm)		0—5	20—30	40—50	0—10	20—30
Particle size distribution (%)	2·0—0·2 mm	57·4	45·4	48·0	13·1	14·0
	0·2—0·02 mm	25·7	26·1	24·0	61·9	52·2
	0·02—0·002 mm	2·5	3·0	5·0	7·8	6·1
	<0·002 mm	11·5	15·0	21·5	15·5	26·6
pH (water)		4·8	4·4	4·4	6·3	6·0
Org. C (%)		1·27	0·42	0·34	0·77	0·45
Total N (%)		0·07	0·03	0·03	0·08	0·06
Total P (ppm)		114	—	79	198	119
Exchangeable cations (me/100 g)	Ca	0·80	0·25	0·05	3·23	1·88
	Mg	0·50	0·10	tr	0·59	0·48
	K	0·15	0·05	tr	0·52	0·25
	Na	0·05	tr	tr	0·08	0·07
	Sum	1·50	0·40	0·05	4·42	2·68
C. E. C. (me/100 g)		5·60	3·30	3·15	6·3	4·4

TABLE 10

A Ferrisol profile (Klinkenberg and Higgins, 1968).

Parent material		*Basement Complex*			
Annual rainfall (mm)		1790			
Site		*Northern Nigeria* (7° 20′ N, 11° 6′ E)			
Sample depth (cm)		0—20	20—33	33—48	48—71
Particle size distribution (%)	2·0—0·2 mm	36	34	38	31
	0·2—0·02 mm	30	22	24	25
	0·02—0·002 mm	12	10	8	12
	<0·002 mm	22	34	38	32
pH (water)		6·3	5·6	5·6	5·8
Org. C (%)		3·52	2·08	0·55	0·40
Total P (ppm)		500	420	340	340
Exchangeable cations (me/100 g)	Ca + Mg	3·92	0·96	0·88	1·04
	K	0·46	0·29	0·22	0·16
	Na	0·05	0·05	0·06	0·03
	Sum	4·43	1·30	1·16	1·23
C. E. C. (me/100 g)		17	13	12	12

(c) *Ferrisols* (Table 10)

These soils are transitional between the Tropical Ferruginous and Ferrallitic Soils. Their normal development is retarded by surface erosion which keeps the profile relatively shallow and forces it to develop into the less weathered parent material (d'Hoore, 1964). Like the Ferrallitic Soils, Ferrisols are characteristically forest soils and their occurrence under savanna is almost entirely limited to a broad area over basement complex rocks in Ivory Coast, Guinea and southern Mali, where they are associated with Ferruginous Soils. No literature has been found which describes this area, but it seems probable that the distribution of Ferrisols and Ferruginous Soils is determined primarily by topographical factors.

Ferrisols are generally more fertile than either Ferruginous or Ferrallitic Soils, having a higher content of exchangeable bases, better structure and high biological activity (d'Hoore, 1964). They might be said to combine some of the advantages of Ferruginous Soils on basement complex with those of Ferrallitic Soils; that is, they have a moderately rich weathering parent material within the root zone of vegetation sufficiently vigorous to accumulate nutrients in a humic topsoil. One important limitation on their agricultural development might, however, be the slope of the land.

(d) *Brown and Reddish-Brown Soils of Arid and Semi-Arid Regions* (Table 11)

These soils occur widely where the mean annual rainfall is less than 500 mm. The parent material is often aeolian in origin. Weathering and leaching are slight, so that these soils often contain appreciable amounts of 2:1 lattice clays and, depending on the nature of the parent material, reserves of weatherable minerals. Organic carbon contents are low or very low, but the organic matter is highly humified and well distributed in the profile. Calcium carbonate concretions may be present at depth.

These soils may also occur under rather more humid conditions (d'Hoore, 1964). In this case formation appears to be favoured by base-rich parent materials and impeded drainage. Tomlinson (1965) suggested that they may form in drift material where it is underlain by impermeable rock in such a position that leaching is prevented without causing poor drainage.

Despite the low contents of organic matter and clay in the topsoil, exchange capacities, reflecting the presence of montmorillonite, are often higher than those of Ferruginous Soils. There is a certain amount of internal leaching, but not much movement out of the profile. Base saturation is therefore high.

Soil structure and permeability in the natural state are good, but tend to deteriorate under cultivation (d'Hoore, 1964). In any case the agricultural potential is severely limited by the aridity. The best form of utilization is probably extensive grazing, although some arable crops can be profitable with carefully controlled soil management or irrigation.

TABLE 11

Profiles of Brown and Reddish-Brown Soils of Arid and Semi-Arid Regions

	Profile 1 (*Klinkenberg and Higgins*, 1968)			*Profile* 2 (*Durand*, 1965)		
Parent material	—			*Alluvium*		
Annual rainfall (mm)	870			320		
Site	*N. Nigeria* (11° 45′ N, 8° 50′ E)			*Dieri, Mauritania*		
Sample depth (cm)	0—30	30—61	61—90	0—23	23—50	50—75
Particle size distribution (%) 2·0—0·2 mm	3	3	4	53·9	47·1	65·9
0·2—0·02 mm	83	77	82	38·8	35·3	14·7
0·02—0·002 mm	4	4	2	0·8	3·3	3·9
<0·002 mm	10	16	12	6·5	14·3	15·5
pH (water)	5·8	5·9	6·1	7·1	7·0	7·0
Org. C (%)	0·22	0·24	0·12	0·25	0·19	0·06
Total P (ppm)	100	87	93	—	—	—
Exchangeable cations (me/100 g) Ca }	0·92	2·16	2·00	4·4	6·4	8·0
Mg }				tr	2·9	3·4
K	0·19	0·22	0·15	0·15	0·02	0·15
Na	0·02	0·02	0·03	0·00	0·00	0·00
Sum	1·13	2·40	2·18	4·5	9·3	11·5
C. E. C. (me/100 g)	2·90	4·10	3·90	5·6	10·8	11·6

(e) *Eutrophic Brown Soils*

These soils usually occur only on base-rich parent materials (e.g. basalt, basic gneiss) and so have only a limited distribution within the savanna. Boulet (1968) has described Eutrophic Soils over granite in Upper Volta, but their existence is probably due to unusual topographical or pedological circumstances. 'Sols Rouge' in north Cameroun, described by Martin *et al.* (1966), are found over base-rich metamorphic rocks and resemble Eutrophic Brown Soils quite closely but are distinguished from them by pedologists because of differences in the distribution of their iron compounds. One might expect them to have similar agricultural potential.

Eutrophic Brown Soils are characterised by shallowness of profile, fairly high organic matter content, high base status and an excellent surface structure and permeability (d'Hoore, 1964; Maignien, 1963). They occur over a wide range of mean annual rainfall (700—1700 mm), but always in positions of good internal drainage. They seem to be relatively young soils existing in conditions favourable to rapid soil formation, a paradoxical situation that d'Hoore attributed to rejuvenation of parent material either by importation (e.g. alluvium) or by erosion of the surface at a rate comparable to that at which the underlying rock is being weathered.

There is a tendency for them to change into Ferruginous Soils on well drained sites and into Vertisols (Maignien, 1963) or Halomorphic Soils (Kaloga, 1970) where drainage is impeded.

The nutrient status of Eutrophic Brown Soils is high compared with that of most other savanna soils (see Table 12). Reserves of weatherable minerals are often large, partly decomposed parent material is within reach of the deeper rooting species, and 2:1 lattice minerals form a substantial part of the clay fraction. However, deficiencies of phosphate and potassium have been noted (Boulet, 1968; Kaloga, 1970), reflecting almost certainly the mineral composition of the parent material, and, although soil physical properties are usually good, agricultural utilisation may be hampered in some area by the shallowness and stoniness of the profile (Tomlinson, 1965).

(f) *Vertisols*

These are dark-coloured, cracking clay soils of low permeability and poor internal drainage, found either on calcareous rocks and rocks rich in ferromagnesian minerals (Lithomorphic Vertisols) or in depressions enriched by constituents coming from surrounding higher land (Topographic Vertisols) (d'Hoore, 1964).

The clay fraction contains a high proportion of 2:1 lattice clays. For eight samples of clay from "Vertisols Typiques et Argiles Vertiques Typiques" from Upper Volta, Kaloga (1966) recorded:—

	Montmorillonite	*Kaolinite*	*Illite*
Range, %	40—80	20—50	0—10
Mean, %	63	33	4

TABLE 12

Eutrophic Brown Soil profiles

		Profile 1 (*Leneuf*, 1954a)		*Profile 2* (*Klinkenberg and Higgins*, 1968)	
Parent material		*Birrimian schist*		*Basalt*	
Annual rainfall (mm)		1200		810	
Site		*Diebougou, U.Volta*		*N. Nigeria* (11° 03′ N, 11° 50′ E)	
Sample depth (cm)		0—10	40	0—15	15—35
Particle size distribution (%)	2·0—0·2 mm	5·0	3·0	2	3
	0·2—0·02 mm	13·5	15·3	60	49
	0·02—0·002 mm	17·0	17·2	12	8
	<0·002 mm	52·2	55·2	26	40
pH (water)		6·9	7·0	6·7	6·2
Org. C (%)		1·42	—	2·54	1·40
Total N (%)		0·075	—	0·13	0·09
Total P (ppm)		114	97	327	447
Exchangeable cations (me/100 g)	Ca	17·10	22·30	7·42	6·34
	Mg	11·81	10·96	8·66	9·06
	K	0·17	0·36	0·14	0·10
	Na	0·03	0·45	0·06	0·09
	Sum	29·11	34·07	16·28	15·59
C. E. C. (me/100 g)		—	—	33·20	33·80

The proportion of montmorillonite increased with depth, so that an exchange capacity of 20 me/100 g soil at the surface rose to around 32 in the subsoil. Higher values are not uncommon. The exchange capacity of 70 me/100 g soil, quoted by Klinkenberg and Higgins (1968) for a Nigerian Vertisol (68% clay), suggests a very high montmorillonite content.

Although reserves of weatherable minerals are often high and the exchange complex is well saturated with cations, the nutrient status of Vertisols is rarely ideal. First, the phosphate and potassium content of the parent materials tend to be low, so that total and available phosphate and potassium in the soil itself are sometimes no higher than in other savanna soils. Brammer and Endredy (1954) pointed out the low levels of soluble phosphate in the black earth soils of the Accra plains and quoted average values of only 0·2—0·3 me/100 g for the potassium reserves of these soils. Secondly, most Vertisols contain only quite small amounts of organic matter (0·5—2·0 % C). In consequence, total nitrogen levels are also low, and the annual production of mineral nitrogen is usually not sufficient for more than very modest crop yields. Analyses of two profiles are given in Table 13.

Nevertheless, d'Hoore (1964) rated the agricultural potential, at least of the Lithomorphic Vertisols, as high. Their topographical situation often favours good external drainage and prevents significant accumulation of soluble salts. On the other hand, in the Vertisols of topographic depressions poor external drainage may aggravate the effects of inherently poor internal drainage. Quantin (1965), discussing the hydromorphic Vertisols of the Central African Republic, noted that despite high nutrient potential, soil fertility was limited by unfavourable physical properties, and utilization was made difficult by irregular flooding.

(g) *Hydromorphic Soils*

These may be defined as soils, other than Vertisols and similar soils, whose development and characteristics are influenced by permanent or seasonal waterlogging (d'Hoore, 1964). However, the effect of this influence varies according to the situation and does not itself impose any uniform pattern on soil properties. Hydromorphic Soils have higher organic matter status than adjacent well-drained soils, but apart from this few generalisations can be made. Very much depends on the cause and nature of the hydromorphism.

At one extreme are the groundwater laterites of Ghana (Brammer, 1962). Although formed on upland savanna sites, their drainage is impeded by the horizontal bedding of impervious shales and by the clayey nature of the weathering rock over granite. The consequent rainy-season waterlogging has produced ironpans or strongly mottled clay horizons at shallow depths in the profile. As a result these soils are shallow, poor in nutrients, waterlogged or droughty according to season, and suitable for little more than grazing land.

Groundwater laterites form under relatively humid climatic conditions

TABLE 13

Vertisol profiles.

		Profile 1 (*Leneuf* 1954b) (*Lithomorphic*)		*Profile 2* (*Quantin*, 1965) (*Topographic*)			
Parent material		*Felspathic amphibolite*		*Alluvium*			
Annual rainfall (mm)		800—1300		950			
Site		*Southern Togo*		*Birao, C. Afr. Republic*			
Sample depth (cm)		0—10	25	0—2	2—10	20—40	50—70
Particle size distribution (%)	2·0—0·2 mm	5·1	14·2	0·8	0·4	0·4	1·3
	0·2—0·02 mm	28·3	44·7	5·5	5·5	4·3	20·2
	0·02—0·002 mm	17·5	15·7	8·5	7·5	6·0	13·0
	<0·002 mm	44·0	20·0	75·0	76·5	81·0	60·0
pH (water)		6·2	6·7	5·2	5·1	5·3	5·3
Org. C (%)		1·78	—	2·03	1·87	0·75	0·46
Total N (%)		0·095	—	0·170	0·148	0·074	0·052
Total P (ppm)		163	88	502	—	360	—
Exchangeable cations (me/100 g soil)	Ca	20·53	19·64	13·10	12·10	13·05	8·15
	Mg	14·98	15·77	9·75	9·05	8·50	5·15
	K	0·17	0·11	1·15	0·65	0·50	0·25
	Na	0·32	0·39	0·20	0·20	0·45	0·45
	Sum	36·00	35·91	24·20	22·00	22·50	14·00
C. E. C. (me/100 g)		—	—	34·10	33·85	34·00	23·30

(1000—1200 mm rain per annum), but high rainfall is not an essential precondition of hydromorphism. At the other extreme, Pias (1960) noted the common occurrence of Hydromorphic Soils in eastern Chad, where annual rainfall is about 400 mm. Here seasonal waterlogging occurs as a result of poor permeability and very gentle topography. According to position there are (i) temporarily waterlogged, compact, impermeable soils in flat areas, (ii) alluvial soils along temporary watercourses, and (iii) black tropical clays, tending towards true Vertisols, in depressions. Similar conditions exist in parts of Mauritania, and Audry (1961) has commented on the synthesis of montmorillonite that takes place where solutions of cations accumulate. With increasing aridity these soils become halomorphic.

Another instance of the effect of topography occurs on former dune fields. Pullan (1962b) described grey Hydromorphic Soils formed in the seasonally flooded troughs of the interdune depressions in northern Nigeria. Soil pH values were about 8·0 to 8·5 at the surface and rose as high as 9·6 in middle horizons where there were high concentrations of calcium and sodium salts. The adjacent soils of the dune tops and sides are freely drained and even slightly acid in reaction. In Senegal the "dior" and "dek" soils (dune top and depression soils, respectively) are less extreme examples of a similar situation.

The most important Hydromorphic Soils of the savanna, however, are those of the valley bottoms, along streams and rivers. Their occurrence is widespread but very dispersed, so that they tend to be under-represented on soil maps. Large areas are found in the flood-plains of most of the main rivers. The "fadama" of the Sokoto river in Nigeria, described by Pullan (1961), is 110 km long and up to 8 km broad; and the Hydromorphic Soils of the inland Niger delta cover many thousands of square kilometres in Mali.

Of frequent occurrence are the small areas of seasonally waterlogged soils in small valley bottoms, often supporting dry-season gardens, bananas, sugar-cane, early maize, and, during the rains, paddy rice. The gradual transition down the catena from the red and brown soils of the upper slopes into the grey hydromorphic soils of the bottom lands is a common sight in the savanna. It is illustrated in Table 14 by data from northern Ghana (Adu, 1963). These show a gradual increase in carbon content, C:N ratio, and cation exchange capacity down the slope as clay content and hydromorphism increase, but a more erratic pattern for base saturation and pH.

As in the example chosen, many of the valley bottom soils are slightly organic and have a high but relatively unsaturated base exchange capacity. The great local variability may be attributed, at least in part, to the very variable material on which these soils are formed. Much of it is recent fine colluvium from the valley sides or both fine and coarse alluvium deposited in time of flood. Local variations both in texture and mineralogical composition are to be expected, and these inevitably have an important influence on soil

TABLE 14

Soil catena in northern Ghana* (Adu, 1963)

	Upper slope	*Middle slope*	*Lower slope*	*Flood plain*
Soil series:	*Babile*	*Tanchara*	*Tono*	*Pani*
Soil texture:	s. loam/l. sand	loamy sand	loamy sand	silty clay
Drainage:	mod. good	imperfect	poor	very poor
Soil colour:	pale brown	brownish grey	brownish grey	dark grey
Org. C (%)	0·40	0·47	0·82	2·52
C/N ratio	12·1	15·2	15·2	16·5
CEC (me/100 g)	2·84	3·85	7·10	24·87
% base saturation	64	53	68	55
pH (water)	5·65	3·85	6·15	5·00

*Analyses of composite samples, 0—15 cm.

drainage and therefore on the degree of hydromorphism and the chemical status of the soil. Moreover, minor topographical features assume greater importance in areas of high water-table and, through their effect on the vegetation and soil aeration, also influence the chemical nature of the soil.

The agricultural potential of the hydromorphic valley-bottom soils is usually rated as high or moderately high, or at least higher than that of the adjacent well-drained soils (d'Hoore, 1964; Jenny, 1965; Lamouroux, 1969), but they are complex and little understood and will require much study before this potential can be fully realised.

(h) *Other soils*

No detailed account is given here either of the Weakly Developed Soils nor of the Halomorphic and Calcimorphic Soils. In the case of Weakly Developed Soils, by definition they have been little affected by pedogenetic processes, and the predominant influences on their fertility are parent material, topography and soil depth. Little has been written about them. Halomorphic and Calcimorphic Soils cover only very limited areas within the savanna region and have been omitted for this reason.

Chapter 3

THE PRESENT FARMING PATTERN

Two distinct forms of agriculture may be distinguished, cultivation and pastoralism. Cultivation involves a far greater proportion of the population and predominates in most areas, but in parts of the northern savanna pastoralism interacts with it and has locally a significant effect on the maintenance of soil fertility. This, however, is poorly documented.

This chapter outlines the present pattern of arable farming, the crops grown and the cultivation practices used. The effects of these practices and those of the pastoralist on soil fertility are discussed later, in Chapter 9.

1 AGRICULTURAL SYSTEMS

The following is a brief discussion only. More detailed accounts of common agricultural systems have been given by Nye and Greenland (1960), Allan (1965), Morgan and Pugh (1969) and Ruthenberg (1971).

Cultivation systems are multitudinous, but three main forms may be recognised:—

(i) Shifting cultivation: a more-or-less haphazard movement of cultivation from place to place as fertility is exhausted or weed populations become uncontrollable, often with movement of village sites as well. Shifting cultivation is necessarily associated with a low population density of about 1·5—9 persons per km^2 (Morgan and Pugh, 1969).

(ii) Rotational bush fallowing: a deliberate alternation between cropping and bush regeneration, the duration of each phase of the cycle depending on soil fertility, weed encroachment, and population pressures on the land. Morgan and Pugh (1969) give a range of 2—32 persons per km^2 in the sorghum-growing areas with lower densities to the north and higher densities to the south.

(iii) Continuous cultivation: associated with high population density usually in the vicinity of large towns, as within 48 km radius of Kano where the population density has been estimated to be well in excess of 100 per km^2 (Dix, 1967; Morgan and Pugh, 1969; Mortimore, 1971).

It will be seen that "shifting cultivation" is used here in a more restricted sense than that used by Nye and Greenland (1960), who included "rotational bush fallowing" within the meaning of the term. In reality, there is no sharp distinction between the two; rather, there exists a continuous transition from shifting cultivation through rotational bush fallowing to intensive continuous cultivation using inorganic fertilizers.

Allan (1965) has proposed a land use factor, which is effectively the ratio of the total land requirement of a farmer to the area he cultivates in any one year. Under shifting cultivation, as defined above, this factor has a value of ten or more, under continuous cultivation a value approaching one. But shifting cultivation is rare in the savanna, and continuous cultivation, although of great local importance, covers only a small fraction of the total farmed area. By far the greatest area is under some system of rotational bush fallowing, with a land use factor of between 2 and 6.

Although bush fallowing predominates, even within the domain of one village the intensity with which the land is farmed may vary very considerably according to its quality and distance from the homestead (Nye and Greenland, 1960). Land nearest the homestead may be cultivated continuously as a garden, with all available manure and household refuse being returned to the soil. At a greater distance rotational bush fallowing is practised, with the length of fallow tending to increase with distance from the homestead, until at the extremity the land use pattern resembles shifting cultivation. Allan (1965) described such a system of "concentric ring cultivation" in northern Ghana and likened it to the old infield-outfield system in Scotland. Similar systems have been described by Nye (1951) also in Ghana, in southern Sudan by de Schlippe (1956), and in northern Nigeria (Agricultural Dept., N. Nigeria, 1953).

In all areas population density has an important bearing on the farming system and, where it is high, fallows are necessarily short and the land use factor low. Although many cultivators own some livestock, particularly poultry and goats, mixed farming is virtually unknown in the savanna. Even where pastoralism and cultivation are practised by the same farmer (e.g. settled Fulani) the two activities are little integrated (Allan, 1965).

Only among the Serere of Senegal are cultivation and stockraising co-ordinated (Guilloteau, 1954; Allan, 1965).

Land use practices are closely linked with land tenure systems. Traditionally land was held communally, cultivation rights being decided democratically or by the village head or elders. At the present time rapid social change is breaking down the traditional systems, and economic pressures are increasing the risk of exploitation of cheap land by people who have no long-term stake in its continued productivity. Guilloteau (1954) has reported that in Senegal migrant labour, business employees and civil servants clear and exploit new land, growing largely cash crops for immediate gain. On the other hand, Dennison (1960) concluded that the trend away from communal ownership, through tenancy, towards freehold possession was likely to be beneficial, since ownership would encourage good management and manurial practices.

2 FARMING METHODS

The following features are common to most traditional agricultural systems in the savanna:

(i) Land is cleared by cutting and burning *in situ*. Clearing is rarely complete, trees of economic value being spared. Where a return to bush fallow within a few years is expected, stumps of other trees and shrubs are allowed to remain to facilitate rapid regeneration.

(ii) Tools are few and simple, the most important being the cutlass and the short-handled hoe. Many varieties of hoe are used, the design varying with function (Morgan and Pugh, 1969).

(iii) Most crops are grown on mounds or ridges, the dimensions of which vary widely. The main advantages would appear to be firstly, the creation of a relatively deep rooting zone of fertile topsoil and, secondly, the improvement of soil drainage. Although terracing systems have been described in certain limited areas (White, 1941; Netting, 1968), elsewhere soil ridges are rarely aligned with any regard to topography and may contribute to erosion.

(iv) The manurial value of animal dung and household refuse is usually well appreciated, and the pasturing of cattle on crop-land during the dry season is encouraged. Crop residues that cannot be put to any other use (such as for fodder, roofing, fencing) are burned on the field.

(v) Most farmers practise a crop rotation and, where the length of the rainy season permits, two crops are grown per year.

(vi) Interplanting (mixed cropping) predominates over monocropping. It is described more fully in the next Section.

(vii) Precedence in planting and weeding is usually given to food crops, so that cash crops, where grown, tend to be planted late. Among food crops, those species and crop varieties that give some yield even in the most adverse season are preferred to more erratic higher yielding varieties. Safety first is the rule in subsistence farming.

(viii) In many areas agricultural production is limited less by shortage of land or low fertility than by shortage of labour at peak periods (sowing and weeding), by the lack of a reliable market (particularly for food crops), and by poor communications to the existing markets.

3 INTERCROPPING

Over the broad range of savanna soils and climates, cropping systems and rotations vary widely. Descriptions of some of them have been given by Nye (1951), Agric. Dept., N. Nigeria (1953), Niquex (1959), FAO/ICA (1960), Mc Ewan (1962), Jordan (1964), Allan (1965), Hasle (1965), Norman (1967), Jurion and Henry (1969), and Buntjer (1970).

An essential feature of all of them is the full use to which the land is put during the often short period of cropping between fallows. Where possible, two crops per year are grown, and in most areas some form of intercropping or sequential planting is practised. For example, in the Zaria area of Nigeria farmers rotate both crops and the positions of their crop ridges and group different crop varieties in certain spatial patterns, interplanting the main crops, either sorghum or a millet/sorghum mixture, with supporting crops at lower population densities (Buntjer, 1970; Norman, 1970, 1975). Less than 17 per cent of the total cultivated area is under sole crops, and, while mixtures of two crops are the most common, mixtures of up to six crops have been found (Norman, 1967, 1970). Certain crop combinations are recognised as mutually beneficial, while others are avoided (Buntjer, 1970).

There are several economic advantages of intercropping: increased returns per hectare (Andrews, 1970; 1974), increased returns per man-hour (Norman, 1970), a more efficient use of scarce land and of a short growing season, insurance against the failure of any one crop and the reduction in losses from diseases, insects and weeds. These benefits are generally recognised, though the effects of intercropping on soil fertility appear to have received little or no attention. They are discussed in Chapter 9.

4 CROPS

Throughout the savanna region the most important crops are food crops grown by the farmer mainly for his own consumption. Cash crops have considerable importance in certain areas, for example groundnuts in Senegal, Gambia, and in the Kano area of Nigeria, and also cotton; but few farmers grow solely cash crops. Rather, the farmer grows food crops for his subsistence, but cash crops only according to the time, land and labour available and the expected market price.

Over the greater part of the region cereal crops predominate, mostly millet and sorghum varieties; but in the Southern Guinea zone where the rainy season is longer yams and maize arc the staples. For every crop there are many local varieties, each one adapted to the local conditions. Curtis (1965, 1966) found that in Nigeria local sorghum varieties are adapted to the length of the rainy season, so that their flowering time always coincides approximately with the end of the rains in that location. Many local varieties are adapted to low fertility and fail to respond to fertilizers. On the other hand introduced high-yielding varieties may fail to yield at all without fertilizers. For increased yields, new varieties and fertilizers must be introduced together.

The main crops of the savanna region have been described in some detail by Purseglove (1968; 1972), Morgan and Pugh (1969), Irvine (1969) and Arnon (1972); and Papadakis (1965) has set out their climatic limitations. The following is a summary:—

(i) Millet (*Pennisetum spp*), being a drought tolerant crop, is dominant in the northern Sudan and Sahel zones, where the rainy season is less than five months. Some varieties mature in less than 90 days from planting. It is also grown far south into the Northern Guinca zone, being planted with the first rains to provide an early harvest. Under these conditions it is usually planted at low density and is later interplanted with sorghum.

(ii) Sorghum (*Sorghum vulgare*) is the staple over much of the southern Sudan and Guinea zones. Most varieties mature late (4-8 months), and are tall-stemmed (2·5—6 m). Flowering occurs near the end of the rains and the grain ripens in dry conditions. Indigenous, long-season varieties require more water than millet but less than maize, and are also more resistant to periods of drought than maize. Intercropping particularly with millet and with groundnuts is very common.

(iii) Rice (*Oryza spp*) Swamp varieties are important locally in riverain areas where seasonal flooding provides suitable conditions (e.g. Casamance and Upper Niger rivers). Pluvial (hill) rice is widely grown in western savanna areas (Senegal, Guinea, northern Ivory Coast), but much less widely in the east where it is largely replaced by yams; it is often interplanted with other cereals or cotton.

(iv) Maize (*Zea mays*) is the predominant cereal crop only in the south of the Guinea zone (i.e. in central Ivory Coast and in a belt stretching from Western Nigeria through Dahomey and southern Togo into the narrow strip of coastal savanna in Ghana). Further north maize is widely grown but production is limited mainly to manured land close to the homestead and to valley bottoms. It requires a higher level of soil fertility than the indigenous varieties of sorghum.

(v) Yams (*Dioscorea spp*) are the staple crop, along with maize, in the southern Guinea zone from the Ivory Coast eastwards. The growth period is relatively long (200—300 days), so that a long rainy season (seven months or more) is needed.

(vi) Cassava (*Manihot utilissima*) is widely grown, particularly in the yam belt, but its wide tolerance of soil and moisture conditions makes its cultivation possible even in the Sudan zone. It will grow in soils of low fertility and is frequently planted last in a rotation.

(vii) Groundnuts (*Arachis hypogea*), important as both a food and cash crop, are grown over most of the savanna, but the areas of greatest production are in the Sudan and Sahel zones, particularly in Senegal, Gambia and Nigeria. Due to its fruiting habit sandy soils are preferred. It is tolerant of low fertility but responds to improved supplies of phosphate and sometimes sulphur. Groundnuts are frequently interplanted with cereals. The haulm is used as animal fodder.

(viii) Cotton (*Gossypium hirsutum*) is most successfully grown in the northern Guinea and southern Sudan zones. In areas having a long wet season it may be planted as a second crop following early millet or maize to ensure ripening under dry conditions. In areas having a shorter wet season the late planting of cotton, as a result of the low priority accorded to cash crops, often leads to very low yields. When protected against insect damage it responds well to improvements in soil fertility.

(ix) Cowpea (*Vigna sinensis*): the seeds are an important source of plant protein for human consumption, and the haulm is used for animal feed. It is usually interplanted with millet or sorghum. It often responds to phosphate fertilizer when protected against insects.

In addition to those listed above, the following crops also have considerable importance in certain areas: benniseed (*Sesamum indicum*); soya beans (*Glycine max.*); tobacco (*Nicotiana tabacum*); fonio, acha (*Digitaria exilis*); sweet potato (*Ipomoea batatas*); Bambarra groundnut (*Voandzeia subterranea*); sugar cane (*Saccharum officinarum*); onions (*Allium cepa*); kenaf, rama (*Hibiscus cannabinus*), and a wide variety of vegetable and fruit crops.

Factors affecting choice of crops are complex (Johnston, 1958; Morgan and Pugh, 1969). Clearly, rainfall is important (Papadakis, 1965), but frequently social, historical, economic and, not least, agricultural factors cause crops to be grown outside their normal climatic zone, or limit their production within that zone. Some examples may be given. Climatically maize is suited to all the Guinea Savanna zone. The predominance of sorghum in most of this area is probably a reflection of the high and intense nutrient demand of maize, particularly for nitrogen, which most of the soils in this zone are unable to satisfy. The long-season, native sorghums have the ability to produce some yield under even the most adverse conditions. However, the value of introduced, short-season maize varieties is becoming recognised, and where fertility is higher, in compound land and fadamas, it is grown increasingly throughout both Guinea and Sudan zones. In the future, increases in the use of mineral fertilizers may lead to an encroachment by maize into areas dominated at present by sorghum and sorghum-millet mixtures.

Millet is a crop which by its drought resistance and short growth period is ideally suited to the northern Sudan and Sahel zones. Yet, interplanted with sorghum, it is grown extensively in the more humid Guinea zone. While the association with sorghum may lead to a greater efficiency in the use of soil and climatic resources (Andrews, 1970), the main reason for the widespread cultivation of millet appears to be the value of its early harvest as a hunger-breaker (Guilloteau, 1954; Johnston, 1958). Early sown maize and fonio have a similar function in some areas (Johnston, 1958).

Cassava yields best in the more humid climates of the Forest and Southern Guinea Savanna zones, but it is widely grown in the Northern Guinea and Sudan zones, where despite lower yields its tolerance of poor soils and drought, and its utility as a perennial food store, make its cultivation worthwhile.

Table 15 lists yields and areas of the major savanna crops (FAO, 1971), but before drawing any conclusions from them, their limitations should be borne in mind. First, data are not available for the savanna alone, and those in Table 15 refer to all the ecological zones in each country. The crop areas in the savanna will therefore be less, and yields may differ from the country averages. Secondly, because of the immense difficulties in collecting agricultural data in Africa, these figures must be regarded as approximations only. Thirdly, and perhaps most important, yields will be grossly misleading if they include substantial areas of mixed cropping, but the extent of mixed cropping is not clear in the published data. Nevertheless, the table indicates very low crop yields, and this is supported by similarly low yields on control plots in experiments to be reviewed in Part III.

5 FERTILIZER USE

Subsistence farming with shifting cultivation and bush-fallowing has allowed crops to be grown for millenia, but fertilizers have been used in recent years, especially since the expansion of cash cropping. A recent estimate of fertilizer consumption is given in Table 16. Total consumption and the amount used per hectare (assuming the estimates of the cultivated areas are of the right order) are both very low. Average use of fertilizer is misleading because most is used on cash crops like groundnuts, but on a regional basis it is apparent that fertilizers at present make only a very small contribution to crop production. Because of its great importance under intensified agriculture, further consideration of fertilizer use is left to Part III.

TABLE 15.

Crop production 1969-70, crop areas in 1000 ha, and yields in kg/ha (FAO, 1971).

	Total land area	*Millet and sorghum*		*Maize*		*Paddy rice*		*Groundnuts (in shell)*		*Cowpeas*		*Cotton (lint)*		*Cassava*	
	10^6 ha	*Area*	*Yield*	*Area*	*Yield*	*Area*	*Yield*	*Area*	*Yield*	*Area*	*Yield*	*Area*	*Yield*	*Area*	*Yield*
Central African Rep.	62·3	80	630	58	780	14	1030	105	810	n.d.		126	140	200	5000
Chad	128·4	1050	680	12	1670	33	1000	160	720	n.d.		283	130	17	3200
Dahomey	11·3	120	520	360	600	3	3330	98	660	n.d.		56	350	113	6500
Gambia	1·1	42	1070	1	550	32	1880	150	780	n.d.		n.d.		1	5300
Ghana	23·9	200	500	450	690	55	1270	98	710	n.d.		n.d.		170	9400
Guinea	24·6	260	580	55	1240	300	1330	30	830	n.d.		n.d.		32	15000
Ivory Coast	32·2	85	490	315	760	290	1070	53	810	n.d.		53	280	163	3300
Mali	124·0	1300	690	100	800	171	880	320	530	n.d.		75	330	10	15500
Niger	126·7	2000	450	3	600	16	2500	300	800	920	160	25	150	29	6900
Nigeria	92·4	3400	820	1100	1110	300	1830	1214	910	1400	500	350	110	1100	6600
Senegal	19·6	1050	620	55	910	120	1420	1010	950	80	380	21	460	40	4000
Togo	5·6	300	430	220	450	26	850	45	400	n.d.		90	70	150	7800
Upper Volta	27·4	672	590	67	620	41	900	140	490	300	280	96	160	3	16100

Note: n.d., no data.

TABLE 16

Estimated consumption of fertilizers, 1970/71 (FAO, 1971) (10^3 metric tons of nutrient)

	N	*P*	*K*	*Area** (10^3 *ha*)
Central African Republic	1·2	0·3	0·2	2,900
Chad	1·6	0·7	n.d.	7,000
Dahomey	2·0	0·3	3·3	596
Gambia	0·2	$<$0·1	—	200
Ghana	1·0	0·4	0·4	2,835
Guinea	1·5	0·2	0·8	n.d.
Ivory Coast	7·8	1·1	11·8	1,859
Mali	3·0	1·1	n.d.	11,600
Niger	0·2	n.d.	n.d.	4,177
Nigeria	5·5	3·1	0·5	21,795
Senegal	3·0	1·2	1·4	5,564
Togo	0·2	0·1	$<$0·1	2,160
Upper Volta	0·4	0·1	n.d.	5,377

*Approximate areas (see text) under arable and permanent crops; for Gambia including areas of shifting and bush fallows; n.d., no data.

PART II—FERTILITY AND MANAGEMENT OF SOILS UNDER PRESENT FARMING SYSTEMS

Chapter 4

SOIL PHYSICAL PROPERTIES

The fertility of a soil is determined both by its physical properties and by its nutrient resources, for the efficient utilization of the latter depends on the suitability of the soil as a rooting medium. Moreover, in a region in which annual rainfall is generally less, and often much less, than potential evapotranspiration, water supply is frequently the factor that most limits crop growth, and the moisture storage characteristics of a soil are an important factor in its productivity. The highly seasonal nature of the rainfall means that periods of water excess may be expected even in the drier areas, and the risks of serious losses of soil and nutrients through erosion and leaching are high. An understanding therefore of the physical properties of the soils, particularly with regard to water control, is essential to the effective utilization and conservation of the soil resources of the region.

This chapter looks briefly at soil physical properties and their interactions with climate, with particular reference to erosion.

1 SOIL PROPERTIES

(a) *Texture*

Over most of the savanna, soils are predominantly sandy, at least at the surface, and this fact underlies many of the problems of moisture control and soil management in the region. Exceptions are provided by the Vertisols and also by some Hydromorphic and Eutrophic Brown Soils, in which clay contents may be quite high, but the main soil types all have mean topsoil clay contents of about 6 to 10 per cent (Table 7). Where values are markedly higher than this, one may suspect profile truncation, for clay generally increases with depth.

Reasons given for the sandy nature of the topsoils vary. The widespread occurrence of sedimentary, granitic and gneissic parent materials is certainly an important factor, but downward eluviation of clay and, possibly, the chemical destruction of kaolinite in the topsoil are also involved (Ahn, 1968; Charreau and Nicou, 1971a); and the erosion of cultivated soils, which is common, also removes a proportionately greater amount of silt and clay than sand (Bertrand, 1967a, b; Kowal, 1970c; Siband, 1972).

Although topsoils are usually sandy, the sand is often largely within the fine sand fraction (0·20—0·02 mm). It is this feature, particularly where there is some silt present as well, that appears to be responsible for the widespread tendency for soil surfaces to form a cap (or crust) after rain.

Stoniness is rare, except in young and weakly developed soils in areas of more rugged topography, e.g. Eutrophic Brown Soils (Tomlinson, 1965); but there are also large areas in which the soils contain considerable quantities of ironstone fragments and ferruginous concretions, derived from the erosion and reworking of ironstone outcrops and ferruginous subsoils.

Increase in clay content with depth appears to be related to climate. Thus the most pronounced textural B horizons are found in leached Ferruginous Soils in the more humid parts of the savanna. The phenomenon is much less marked in non-leached Ferruginous and Brown Sub-Arid Soils (Table 11) and in the forest zone. Although eluviation of clay and iron out of the topsoil is a characteristic of leached Ferruginous Soils (d'Hoore, 1964), the degree and depth of clay accumulation depend on parent material and topography as well as on climate. In the extreme, the soils develop into groundwater laterites, in which clay and iron pans may occur quite close to the surface, seriously limiting water movement and crop rooting (Adu, 1972). In Ferrallitic Soils the increase in clay with depth is more gradual, and subsoil structure is more favourable.

(b) *Structure*

Many surface soils contain only weak aggregates, a feature that is probably related to the low clay content and to the predominance of kaolin in the clay fraction. Subsoils are generally compact and massive and constitute a severe limitation to rooting at many sites (Nicou *et al.*, 1970; Bertrand, 1971; Haddad and Seguy, 1972). To give one example, Berger (1964), in Ivory Coast, noticed that yam elongation ceased when the tuber reached the base of the loosened mounded soil, and the tap root of cotton failed to extend into the subsoil unless this had been loosened when the crop ridges were made.

Surface soils under natural woodland in more humid areas may, through mesofaunal activity on the litter, have a good porous structure, but this is rapidly lost on cultivation (Charreau and Tourte, 1967). Commonly, cultivation produces at best only unstable clods with little or no resistance to heavy rain. When, after a few storms, these have been broken down, the smoothed soil surface dries as a crust, which therefore restricts infiltration and aeration and increases runoff (Lawes, 1961; Evelyn and Thornton, 1964; Kowal, 1968a). It may also restrict seedling emergence. When the crust is broken by further cultivation, the soil structure tends to become single-grained, and the aggregation even more ephemeral as cultivation is continued (Martin, 1963; Siband, 1972). Pale, dusty soils are a common sight on fields that have been farmed for many years, and in more arid areas these soils bear a high risk of serious wind erosion.

Soil structure has proved difficult to quantify, although certain properties related to structure, such as porosity, may be fairly easily measured. Francophone workers have found it useful to estimate the stability of the soil structure in terms of Henin's index of structural instability (Henin *et al.*, 1969):

$$I_s = (A + L) \div \frac{(Ag' + Ag'' + Ag''')}{3} \text{-} 0{\cdot}9\ Sg$$

where (A+L)=percent of silt plus clay following maximum dispersion, Ag', Ag'', and Ag''' are the percentages of aggregates greater than 0·2 mm retained on wet sieving after pretreatments: nil, with benzene, and with alcohol respectively, and Sg is the percentage of sand greater than 0·02 mm. Under savanna conditions this parameter increases under cropping and decreases under fallow, and it appears to reflect quite closely the general physical status of the soil. However, Charreau and Nicou (1971a) reported that the method presents difficulties when the sand content is high, and they suggested that the coefficient of infiltration, K, is more valuable for characterizing the structure of sandy savanna soils.

Two cohesive factors are involved in aggregation in these soils. The importance of one of them, organic matter, is referred to in Chapter 5, where it is noted that structural stability, measured in terms of Henin's index, I_s, correlated strongly with the content of non-humified organic matter in some soils of the Central African Republic (Combeau and Quantin, 1964). A similar conclusion was reached by Martin (1963), working on savanna soils in the Niari valley in the Congo Republic, but he also drew attention to the second factor, the free iron oxides. At this site, particularly in cultivated soils, free iron oxide, as a percentage of clay, correlated well with structural stability.

Russell (1968b) made a distinction between soils derived from rocks low in iron (sandstones, granites, etc) and those high in iron (e.g. basalts), suggesting that whereas the former tend to be rather weakly structured, in the latter the iron oxides bind soil particles into stable crumbs. Other authors have seen the main difference as lying between soil types or, more particularly, between the different physicochemical processes involving iron oxides in the two main soil types: in Ferruginous Soils iron oxides are either leached out of the profile or precipitated as mottles or concretions; in Ferrallitic Soils they are dispersed through the profile in association with clay surfaces (d'Hoore, 1964).

At Sefa, Casamance, the presence of "pseudo-sand" aggregates (cemented by ferric oxides) in a weakly Ferrallitic Soil gave it a better and more stable structure than that found in a leached Ferruginous Soil formed over similar parent material nearby (Charreau and Fauck, 1970; Charreau and Nicou, 1971a). Similarly Ahn (1968) showed that much of the clay fraction of several Ghanaian subsoils was aggregated into silt-sized, even fine sand-sized particles. As a result, the effective texture of these horizons was much sandier than that indicated by particle-size analyses using chemical dispersing agents, and although there was little apparent macrostructure, physical properties, including percolation, were good and stable.

As has been said, structure deteriorates quite rapidly with cultivation. At Grimari, Central African Republic, the instability index, I_s, increased from an initial value of 0·4 under savanna vegetation to 1·9 after eight years' cropping (Morel and Quantin, 1972). At Sefa, I_s increased from 0·4—0·7 under woodland to 1·3—1·6 in fields cropped for six years, the increase being greater in Ferruginous than in Ferrallitic Soil (Charreau and Fauck, 1970). However, in so far as the deterioration in structure is due to a loss of organic bonding through enhanced mineralization of organic matter, the problem is common to all types of soil.

(c) *Soil hardness*

Almost all savanna soils undergo a very marked hardening during the course of the dry season. From the evidence of penetrometry studies in Senegal, summarized by Charreau and Nicou (1971a), the degree of hardness appears related to clay content. Even the very sandy 'dior' soil became hard on drying, but its resistance to penetration was only about half that of the 'dek' soil or the leached Ferruginous Soil at Sefa. Hardness increased with depth, particularly at Sefa, because of increasing clay content.

The mechanisms involved in hardening are not understood. At any particular depth hardening does not start until about half the moisture content at field capacity has been lost, but thereafter it increases with time for as long as evaporation continues.

The hardness of the dry season soil is a serious constraint on land management. It makes cultivation almost impossible in many areas throughout the dry season, with the consequence that the penetration of the early rains into the hard crusted soil is often far from complete and, since cultivation must await the moistening of the topsoil, the effective length of the growing season is appreciably reduced. As discussed in Chapter 11, ploughing at the end of the cropping season can in some circumstances be used to overcome this problem.

(d) *Porosity and permeability*

Charreau and Nicou (1971a) remarked on the high bulk density (apparent density) frequently found in surface soils, with values as high as 1·6—1·7 in cropped, but untilled fields, implying porosities of only 36—40 per cent. Topsoil porosities at Samaru, reported by Kowal (1968 a, b), were rather higher, 41—46 per cent, but fell to about 36 per cent in the subsoil.

The values vary with cultivation history. Tillage creates high surface porosity, but the aggregates formed are unstable, and exposure to rain quickly leads to the formation of a surface cap. Macroporosity is much reduced, with serious consequences on the infiltration rate. However, at this stage the surface cap, only a few millimetres thick, is easily broken by light hoeing, and infiltration can be easily, if only temporarily, restored. On the other hand, where land has been cropped for several years with only the most superficial cultivation, the whole topsoil may be compacted, with reduced macroporosity and permeability throughout. Data from a Samaru plot, bare-fallowed for several years (Kowal, 1968a), illustrated the difference in pore-size distribution between a recently tilled and a compact surface soil:

	Bare-fallowed soil	*Tilled soil*
Total porosity, %	41·6	47·2
Macroporosity, %	10·3	21·8

Studies in Senegal have shown the serious effect that low porosity can have on root growth and crop yield. Charreau and Nicou (1971a) reported that sorghum yields decreased linearly by 450—600 kg/ha for every 0·10 increase in apparent soil density. They thought this was due more to mechanical resistance to root extension than to poor aeration, for water-free pore space was rarely less than 10 per cent, which is generally thought adequate for root aeration. Nevertheless the risk of inadequate aeration exists in these compact surface soils, particularly at the height of the rains, but the problem has so far received little attention.

Measured values of infiltration and permeability vary widely according to the method adopted. Charreau and Nicou (1971a) reported that although permeability measurements on sieved samples in the laboratory gave rather low values, field methods gave 20—50 cm/h at Bambey and 10—20 cm/h at Sefa. These figures did not accord with rainy season experience, for the ability of the soil to absorb high rainfall intensities decreased as the season progressed. Only an indirect method based on runoff rate as a function of time gives a true picture of rainfall acceptance under field conditions. This is because the existing soil moisture status is critical; at Sefa infiltration fell during the course of one storm from 5·0 to 0·5 cm/h.

Subsoil permeability is also important. During periods of heavy rain the depth and permeability of the least permeable horizon may control conditions in the topsoil. Measurements at one site at Sefa showed the permeability of the 20—100 cm horizon to be lower than that of the top 20 cm by a factor of thirty, a result which accounted for the waterlogging often observed after heavy rain (Charreau and Nicou, 1971a). In extreme cases, the perched water-table may rise to the surface and cause sheet flow (Adu, 1972).

Low subsoil permeability is a common feature of leached Ferruginous Soils, because of the accumulation of clay at depth; but where sesquioxide-bound aggregates are abundant, as in some Ferrallitic Soils, subsoil permeability may be satisfactory despite the high clay content. The soils described by Ahn (1968) had high subsoil clay contents and little obvious macrostructure, but water passed through them rapidly. Percolation is also relatively unimpeded through non-leached Ferruginous Soils and Sub-Arid Brown Soils, which have little accumulation of clay. Figures of Charreau and Nicou (1971a) showed that rainfall acceptance at Bambey (non-leached Ferruginous Soil) was markedly higher than that of the leached Ferruginous Soil at Sefa.

(e) *Available water holding capacity*

In many of the drier areas of the savanna, water supply is the first limiting factor on crop production and on the effectiveness of fertilizers (Gautreau, 1965; Boulet, 1968; Poulain and Arrivets, 1972). Even in the more humid areas, periods of drought may occur both early and late in the season, with serious effects on yields. The effect is, however, buffered by water stored in the soil.

Within the limits of the sandy surface soils of the region, the greater the clay content the greater the moisture storage (Table 17). Even the small textural differences between the dior and dek soils of Senegal is significant. Charreau and Poulain (1963; 1964) reported that in a dior soil with 125 mm available water in the rooting zone maximum yields of millet were only 2 tonnes per ha, whereas in a dek soil with 210 mm water storage they reached 3·3 tonnes.

Moisture storage is also affected by soil organic matter content, although within the rather narrow range of organic matter levels occurring in well-drained savanna soils differences due to this cause are not likely to be very large. However, Siband (1972) found that the soil moisture equivalent (described as measured at pF 2·5) followed the organic matter content in decreasing with the length of time a soil had been under cultivation. The oldest fields had least available moisture.

The main conclusion of the Senegal work was that available water holding capacity is generally low, ranging from 4 to 8 per cent by weight (Charreau and Nicou, 1971a). Unfortunately, such low values are common over much of the drier savanna. The higher values of Kowal (Table 17) are perhaps typical of Ferruginous Soils with textural B horizons, as occur in more humid areas. One feature of Kowal's results is that much of the available water appears to be loosely held, 75 per cent being released at tensions of less than pF 3·0. Slow drainage therefore continues for several days after a rain storm (Kowal, 1968a). As a result, the conditions under which field capacity is measured may greatly affect the value obtained, and this may account for some of the difference in values between the soils of Samaru and Sefa (Table 17).

(f) *Runoff*

Except where the land surface is completely flat, runoff occurs whenever rain falls more rapidly than the soil can absorb it. The important factors, therefore, are rainfall intensity and maximum infiltration rate. As has been seen (Chapter 1), rainfall intensities in heavy storms may often approach and even exceed 100 mm/h, and there are few soils that can accept rates

TABLE 17

Some physical constants and water-storage characteristics of soils (0—20 cm) at three savanna sites.

Site	Soil type	*Mechanical analysis, %*			*Moisture data, % by weight*		
		Clay	*Silt*	*Fine Sand*	*Field capacity*	*Wilting point*	*Available water*
[1]Bambey	Dior	4	0·5	75	5—6	1·5—2·0	3·5
[1]Bambey	Dek	10	3·5	65	8—9	3—4	4—6
[1]Sefa	Ferruginous	12	3·1	50	9—13	3—5	6—8
[2]Samaru	Ferruginous	13	8	71	18—21	4—4·5	14—71

[1] from Charreau (1965) and Charreau and Nicou (1971a): 'Dior' is a slightly leached Ferruginous soil, 'Dek', a soil transitional between a Ferruginous Soil and a Vertisol; the Sefa soil is leached Ferruginous.
[2] calculated from Kowal (1968a); Samaru soil is leached Ferruginous.

as high as this. Except where the surface is deeply covered with mulch or natural litter, some runoff is inevitable. Cocheme and Franquin (1967a, b) suggested that runoff begins to occur wherever mean annual rainfall exceeds 500 mm, although mean annual rainfall would not seem to be the most relevant parameter in this context.

Measurement of infiltration rate or rainfall acceptance was discussed briefly above. Meaningful values are difficult to obtain, for field infiltrometer and laboratory values are rarely matched by performance under actual rainfall conditions. Lawes (1961; 1964), using small catchment gauges at Samaru, recorded infiltration rates that varied from more than 120 mm/h under mulch or good grass cover to less than 10 mm/h for undisturbed bare soil. He considered that in the latter case, over 70 per cent of the season's rainfall could be lost by runoff. However, this work was conducted on a small scale and took no account of gradient, length of slope, vegetation cover or land management.

The data of Kowal (1970b) from five cultivated plots at Samaru, each 8×183 m^2, set along a falling contour of 1:316 slope, showed over five years:—

Mean annual rainfall:	1062 mm
Mean annual runoff:	140-297 mm (13·2—27·9%)

Runoff varied with management, being greatest from cropped plots ridged along the falling contour (27·9%) and from bare-fallowed land (25·2%), and least from cropped land with alternately tied ridges (13·2%). There was no relation between total seasonal rainfall and total seasonal runoff. On average, 32 storms out of 85 days of rain produced runoff; and, except early in the season, all storms that gave more than 20 mm in 24 hours resulted in some runoff. On two additional plots kept under natural vegetation, burned and unburned, runoff occurred only during a few very heavy storms.

Lawes was right to emphasize the importance of the condition of the soil surface. If runoff is to be minimized, the surface soil must be kept as permeable as possible by frequent cultivation. However, once the profile is wet, surface permeability is not the only factor controlling infiltration; the depth and permeability of the least permeable horizon are also important.

TABLE 18

Effect of slope on runoff and erosion at Sefa, Senegal (seven-year means) (Roose, 1967).

	Slope, % 1·25	1·50	2·00
Annual rainfall, mm	1,235	1,235	1,186
Runoff, %	16·3	21·9	30·0
Erosion loss, t/ha	4·75	8·62	11·81

During periods of heavy rain, a relatively impermeable horizon in the subsoil, as occurs in many leached Ferruginous Soils, can hold up water in the topsoil and give rise to surface flow. It is this phenomenon, as well as increased capping, which causes the higher percentage runoff often observed during the second half of the season (Kowal, 1970b).

Runoff is also a function of slope, the number of years of cultivation and the cultivation techniques used. Seven-year mean values from the runoff plots at Sefa, Senegal, showed that there was almost twice as much runoff from a slope of 2 per cent as from one of 1·25 per cent (Table 18). At the same site, runoff increased with the number of years of cultivation; for the same sequence of crops mean runoff in the years 1960-3 was 35·8 per cent as compared with 17·1 per cent in 1955-8 (Table 19). Roose attributed this difference to a deterioration in soil structure arising from the mineralization of much of the soil organic matter.

TABLE 19

Changes in runoff and erosion with time at Sefa, Senegal (Roose, 1967).

Crop	*Year*	*Rain, mm*	*Runoff, %*	*Erosion, t/ha*
Groundnuts	1955	1340	18·0	16·30
	1960	1302	28·9	6·99
Rice	1956	1148	21·3	6·47
	1961	1185	36·4	10·84
Groundnuts	1957	1004	12·3	6·72
	1962	1223	38·8	3·32
Rice	1958	1484	16·2	18·39
	1963	1279	39·3	9·71
Mean	1st cycle	1244	17·1	11·97
	2nd cycle	1247	35·8	7·72

Note: all results are from the same plot, unreplicated.

The effect of land management and cultivation techniques on runoff and erosion is complex. Roose's figures show that in the early years at Sefa, mechanical cultivation increased runoff compared with traditional hoe cultivation, and a similar result was reported by Verney and Williame (1965) from Dahomey. However, with an improvement in techniques, the position at Sefa has reversed; ploughing at the end of the season decreased runoff in the subsequent year by 44 per cent (Charreau, 1969; Charreau and Fauck, 1970). It seems probable that frequent, rather shallow mechanical cultivations destroy surface structure more rapidly than hand cultivations, but less frequent, deep ploughing, particularly when it involves the incorporation of organic residues, effects a more enduring improvement in these properties. We shall return to this subject in Chapter 11.

2 EROSION

(a) *The occurrence of erosion*

The extent of erosion in the savanna is not known. Gullying is common in some areas; wind erosion is sufficiently serious for windbreaks of trees to be planted in parts of the north; and to judge from the high content of suspended solids in the rivers, sheet erosion (see footnote) occurs widely. Even so, lack of extensive survey data makes it difficult to assess the seriousness of erosion in its various forms in the region as a whole.

A few local surveys have been made. Ofomata (1964; 1965) and Floyd (1965) described catastrophic gullying in the derived savanna of eastern Nigeria, due apparently to deforestation exposing slopes previously stabilized by good cover; in the Sokoto-Rima and Chalawa catchment areas of northern Nigeria appraisal of air-photographs showed that 20-50 per cent of upland soils were gullied (Chalk, 1963); and FAO (1965) estimated that 70 per cent of the soil of northern Nigeria was eroded to some extent. Brammer (1962) considered that erosion was not a serious problem in Ghana, although it occurred in northern areas with high population densities. More recently Adu (1972) described erosion in the Navrongo-Bawku area (9230 km^2) of northern Ghana and stated that sheet erosion was more common than gullying. About 40 per cent of the land was affected by moderate sheet erosion involving some truncation of the profile; another 8 per cent had lost both A and B horizons.

The generally serious nature of erosion found in these local surveys, backed by the opinions of reliable observers (Stamp, 1953; Fournier, 1967; Morgan and Pugh, 1969), makes it seem very probable that erosion is already a major problem through much of the region. Moreover, it is likely to increase, for of the two main causative factors, climate and human activity, the former may remain approximately constant but with expanding population the latter seems certain to intensify.

Fournier (1960; 1962) used a broad climatic approach to map the risk of soil erosion by water in Africa south of the Sahara. Combining an empirical rainfall factor, p^2/P (where p is the rainfall in mm in the wettest month and P the annual rainfall) with an average relief factor for each river basin, he showed that the erosion potential in the savanna was high, particularly in the west. The limitations of his early approach were that the scale could not accommodate local variations in slope and, though climate and slope are important, other factors, like soil properties, length of slope, vegetation and land use, were not allowed for.

More recently, Fournier (1967) and others (Roose, 1967; Charreau, 1968) have adopted for erosion studies in francophone Africa the model of Wischmeier and Smith (1960), in which soil loss is expressed as a function

By sheet erosion we mean the washing of surface soil from land by water flowing over the surface, initiated by the splash of raindrop impact. Use of the term was criticised by Hudson (1971) on the grounds that it implies a flow of uniform depth, which is usually not so. But the combined effert of splash and runoff often gives a more or less uniform removal of soil without any visible rill formation, and this seems to be appropriately described as sheet erosion.

of five separate indices, for rain, soil, slope, crop and conservation system. Results from this approach have been reported from Ivory Coast by Bertrand (1967a) and Roose (1973).

(b) *Experimental work*

When water erodes soil, the energy needed to detach and remove soil particles comes from two different sources, the kinetic energy of falling raindrops and the kinetic energy of flowing runoff water. It seems to be generally agreed that the former is the more important cause of erosion (Fournier, 1967; Kowal, 1970b). The energy dissipated through raindrop impact in heavy storms can be very great indeed, and most of it is expended in breaking up soil aggregates, detaching soil particles and splashing them away from the points of impact. Figures from Niangoloko, Upper Volta, showed that particles could be splashed as far as 120 cm, and the total amount of material transported by splash at this site in 1957 was equivalent to 170 t/ha (Fournier, 1967).

The scouring capacity of running water increases as about the fifth power of its velocity. Thus, while a large volume of water flowing slowly may take very little soil with it, the same volume confined to a narrow channel causes rapid gullying. This may bring catastrophic losses where, for instance, crop ridges are aligned down the slope, but over the region as a whole this is probably a less serious problem than the less spectacular sheet erosion. What initiates the loss in sheet erosion is the dispersion of the soil particles by raindrop impact. Once dispersed the finer particles are easily removed even by slowly moving runoff water.

This explains why the relationship between runoff volume and erosion loss is only a loose one. Both the extent to which the soil is protected from the direct impact of rainfall and the degree to which the runoff is channelled, for example by mounds or ridges, affects the load carried by the water. The importance of these two mechanisms of erosion, splash and scour, in any particular storm is controlled by local factors, in particular rainfall intensity, soil texture, slope, crop cover and management. The first four of these are discussed here; management and soil conservation are left to Chapter 11.

(i) *Rainfall intensity*

Although there is much variability, median drop size, as well as the number of drops, increases with increasing rainfall intensity. The terminal velocity of larger raindrops is approximately proportional to their radius, while their kinetic energy is proportional to both their mass and the square of their velocity. It follows, therefore, that the kinetic energy of rainfall increases very rapidly with increasing intensity, as discussed in some detail by Fournier (1967). His figures for Niangoloko, Upper Volta, showed that falls with a maximum intensity of less than 90 mm/h were very rarely erosive, but erosion was frequent with greater intensities and always occurred when the maximum intensity exceeded 120 mm/h. Losses from storms in this last category accounted for 88 per cent of all the erosion during 1956-7, although the total period of time during which these losses occurred amounted to only 20 hours in the two years put together.

(ii) *Soil texture and structure*

Erosion losses may be related to texture and structure in two ways. Firstly, through permeability, these properties affect infiltration and hence runoff. Secondly, since the magnitudes of the attractive forces between sands and clays differ, the relative importance of splash and scour vary according to which particles predominate in the soil surface. Following Fournier (1967) and Roose (1967), we may say that, in a sandy soil of largely monoparticulate structure the material is easily detached and erosion losses depend on the transport capacity of the runoff water, whereas when the soil contains well-aggregated clay dispersion is less easy, and the amount of material detached and removed depends largely on the violence of the rainfall.

(iii) *Slope*

It is necessary to distinguish between the inclination and the length of a slope. The steeper the slope, the greater the runoff and the rate of flow of runoff, and hence the greater the erosion (Table 18). But also, the longer the slope, the greater the volume of runoff at the lower end, with the probability of increased scour.

(iv) *Crop cover*

The crop canopy absorbs the kinetic energy of the raindrops in proportion to its degree of cover of the soil surface. Subsequent leaf drip and stem flow, because of their lower velocity, are much less destructive. Thus where cover is complete, dispersion of soil particles by rainfall is reduced to a very low level, and runoff, providing it is slow-flowing, carries little load. Because of different growth habit, crops vary in the amount of cover they provide and the rapidity with which it is established, and this may affect the extent of erosion (Table 20). But this factor is less important than early crop establishment. Management systems which facilitate early planting usually show less erosion than those which do not, whatever the crop (Charreau and Fauck, 1970).

As may be seen from Tables 18, 19 and 20, erosion losses from experimental plots fall in the range, 0—30 t/ha, the lowest values coming from fallow and natural regeneration, the highest from plots carrying a failed crop which affords little or no cover to the soil. All these results are from well-managed sites of slope not exceeding 3°. Losses from farmers' fields on steeper, and possibly longer, slopes may be locally much more severe.

Kowal (1970c) calculated that the mean erosion on the runoff plots at Samaru, 10·8 t/ha/annum, represented a loss of 1 centimetre in 15 years. Such a loss may be acceptable where the soil is deep, but many soils overlie relatively soft, sesquioxide-rich horizons, which on exposure may harden irreversibly (Ahn, 1970). There is also a qualitative aspect. In general, material removed by erosion is richer in clay, silt and organic matter than

TABLE 20

Effect of vegetation cover on runoff and erosion at Sefa, Senegal (Roose, 1967).

	Replicates	*Rainfall (P), mm*	*Runoff, mm*	*Runoff, %P*	*Erosion, t/ha*
Natural bush vegetation—burnt	9	1167	11	0·9	0·2
unburnt	7	1138	8	0·7	0·1
Fallow (with some vegetation cover)	7	1203	200	16·6	4·9
Crops—					
groundnuts	24	1241	282	22·8	6·9
cotton	3	1151	322	28·0	7·8
sorghum (green crop)	9	1157	239	20·6	7·8
sorghum (grain)	2	1113	379	34·1	8·4
rice	11	1129	307	27·2	8·2
maize	1	1279	395	30·9	10·3
millet	2	993	345	34·7	10·3
Crop failures	7	1201	482	40·1	25·1

the bulk of the topsoil. For instance, Pieri (1967b) reported that at Sefa the eroded material had a fine fraction four times larger than that of the original soil, and in Dahomey there was a more than twofold enrichment in organic matter (Verney and Williame, 1965). There was a similar pattern at Samaru and Niangoloko (Table 21). Such reports are consistent with observations that field soils become coarser with increasing length of time under cultivation (Bouyer, 1959; Siband, 1972). The actual impoverishment of a site through erosion is therefore usually greater than indicated by the depth of soil lost.

TABLE 21

Granulometric comparisons of eroded material with original soil.

Site and year: *Details:*	1. Niangoloko, 1956 Mean of 3 cultivated treatments	2. Samaru, 1965-8 Mean of 5 treatments (4 cultivated)
Eroded material, %		
Sand	67·1	63·3
Silt	15·2	36·6 (Silt + Clay + humus)
Clay + humus	16·2	
Original soil, %		
Sand	92	80
Silt	5	6
Clay + humus	4	15

Sources: 1. Fournier, 1967. 2. Kowal, 1970c.

However, at any one site the composition of the eroded material varies widely according to the conditions. Bertrand (1967b) reported that the eroded material from the runoff plots at Bouake, Ivory Coast, was coarser when erosion was heavy and generally became coarser as the season progressed; moreover, its composition changed sharply in the third year of cultivation, with less silt and more sand being removed. This may reflect a change in the composition of the surface soil through previous erosion. Further, it is clear from the results of Fournier (1967) and Kowal (1970c) that management practices which reduce total erosion loss may effect a greater conservation of the coarse fraction than of the fine. This is because slow moving runoff water can carry clay but not sand. The composition of the eroded soil is therefore less a function of the nature of the original soil than of past and present land use and management practices.

Runoff and erosion inevitably involve the loss of soil nutrients. At Bouake a loss of 55 tonnes soil per ha over 5 years included 1200 kg organic matter, 75 kg nitrogen, 16 kg phosphorus and 34 kg exchangeable bases, and these amounts do not include the nutrients dissolved in the runoff water (Bertrand, 1967b). The nutrients in the runoff water may comprise a large proportion of the total loss. Mean annual removal of nutrients from the

Samaru runoff plots over four years was, according to treatment, 16—34 kg cations (Ca, Mg, K and Na) per ha and 8—21 kg nitrogen per ha. Of the cations removed more than two-thirds were found in the runoff water (Kowal, 1970c). Considered in terms of the total nutrient resources of these soils such losses are by no means negligible. They are considered further in Chapter 9.

(c) *Wind erosion*

Although no data have been found, there is little doubt that wind erosion occurs widely in the northern savanna. Lamaitre (1954) regarded it as a more serious problem than water erosion in Niger and noted that the ash of bush fires was lost as well as the soil. Further south, the violent winds that precede the first rains move a great deal of soil from cultivated land of degraded structure. The planting of windbreaks in some areas attests to the importance that is attached locally to this problem (Lamaitre, 1954; Guiscafre, 1961).

Chapter 5

GENERAL SOIL CHEMICAL PROPERTIES

1 ORGANIC MATTER

(a) *Amounts*

One characteristic of well-drained savanna soils, with important implications for their fertility, is the generally low level of organic matter. On the assumption that soil organic matter contains 50 per cent carbon, then from the data in Chapter 2 (Table 7) the organic matter contents of surface soils (0—15 cm) range from about 0·5 per cent in the Brown and Reddish-Brown Soils of the Sahel zone to about 2 per cent in Vertisols, Eutrophic Brown and Hydromorphic Soils. In the widely occurring Ferruginous soils the mean content of 245 samples was 1·2 per cent with a range of 0·2 to 5·6 per cent.

These amounts are low not only compared with those in soils of temperate regions (Cooke, 1967), but also in comparison with those in soils from the forest zone of West Africa (Nye and Greenland, 1960) and from a range of altitudes in Kenya (Birch and Friend, 1956).

(i) *Effect of Climate*

The factor having most influence is almost certainly climate and, in particular, the mean annual rainfall and its seasonal distribution. Organic matter increases with rainfall, as has been demonstrated by Gavaud (1968) in Niger, over the range 230—875 mm per annum, and by Kadeba (1970) in

Nigeria. Using data drawn from all the savanna region (350—1905 mm rain per annum), Jones (1973) obtained the relationship:

$C\% = 0{\cdot}137 + 0{\cdot}000865X$ ($r = 0{\cdot}514$***), where X = mean annual rainfall (mm).

This indicates a mean increase in soil organic matter of 0·17 per cent per 100 mm increase in mean annual rainfall. The highly significant relationship between soil organic matter content and rainfall may, however, be a reflection more of the length of the rainy season than of the annual rainfall because these two parameters are closely correlated with each other.

The markedly seasonal character of the savanna climate affects the accumulation of organic matter in several different ways. In the first place, the duration of the wet period, through its influence on the nature of the natural vegetation and its annual growth period, very largely determines total dry matter production and, hence, the quantity of plant residues returned to the soil. Secondly, the length and severity of the dry season determines the risk of bush-fires and their probable destructiveness and therefore, indirectly, the return of residues to the soil. Hopkins (1966) observed that more leaf litter was destroyed by fire than by decay, and on a site in the derived savanna with a relatively short dry season Nye (1959) attributed the slow rate of humus increase to grass-burning and the destruction of litter. One effect of fire is to encourage grasses at the expense of trees and bushes (Hopkins, 1965b; West, 1965), particularly if the burn is late and fierce (Trapnell, 1959; Hopkins, 1965b), and so to alter the amount and nature of organic residues returned to the soil and the distribution of roots through the profile.

Thirdly, the alternation of wet and dry seasons enforces a seasonal pattern on microbiological activity. It might be thought that the shortness of the wet season would effect a relative conservation of organic matter by seriously curtailing the period during which decomposition can occur, but this appears not to be so. Rather the intense drying of the soil during the prolonged dry season leads to a very rapid flush of decomposition on subsequent rewetting. The results of Birch (1958, 1959) and Enwezor (1967) indicate that the longer and hotter the dry season, the greater the percentage mineralization at the start of the next wet season. Moreover, Birch (1958) has shown that the pattern of the flush "is repetitive and the frequency with which it occurs will be an important factor in the rundown of soil carbon". Further, since low annual rainfall "is usually associated with a greater frequency of wet and dry periods than high rainfall it should, above a certain lower limit, be conducive to an accelerated loss of soil carbon", and this might be expected to lead to a direct relationship between total rainfall and the carbon content of the soil.

Two other environmental factors, altitude and the mean annual temperature, which have been shown to affect soil organic matter status in other parts of the world (Birch and Friend, 1956; Harradine and Jenny, 1958; Jenny and Raychaudhuri, 1960) appear to be of little importance in the savanna, where altitude and temperature differences are small. An increase in soil organic matter with altitude has been shown in data from West Cameroun (Hawkins and Brunt, 1965; Jones, 1973), but this area is not typical of the savanna region as a whole.

(ii) *Effect of Parent Material*

A second factor affecting soil organic matter status is the nature of the soil parent material, for this may determine both the texture of the soil, in particular the size and nature of its clay fraction, and also its nutrient status.

The ability of clay to adsorb humus is generally accepted (Birch and Friend, 1965; Stevenson, 1965; Kononova, 1966; Russell, 1973), although the nature of the association between them is still not well understood (Greenland, 1965). The dependence of the organic matter contents of savanna soils on texture was noted by Fauck (1960), Thomann (1964), Gavaud (1968), and Kadeba (1970), and for a wide range of well-drained savanna soils Jones (1973) obtained the relationship:

$$C\% = 0{\cdot}341 + 0{\cdot}0273\ Y\ (r = 0{\cdot}540^{***}),\ \text{where } Y = \text{per cent clay.}$$

Thus for a mean annual rainfall of 1000 mm per annum, the mean soil organic matter content increases from 0·75 to 2·60 per cent over the clay range 0—50 per cent. As already described in Chapter 2, many savanna soils are developed on basement complex rocks and sandstones, both of which tend to give rise to rather coarse-textured soils; in addition, the prevailing pedological processes in these soils have often caused clay eluviation out of the topsoil. Most savanna topsoils therefore tend to be low in both clay and organic matter.

Soil nutrient status also depends on parent material (Chapter 2). Fauck (1963) pointed out the poverty in exchangeable bases of those leached Ferruginous Soils formed over sandstones; and Wild (1971), in Nigeria, showed that soils derived from basement complex had a much higher total potassium content than those derived from sandstone. Similar differences probably occur with other nutrients, and since soil nutrient status seems bound to affect both the nature and vigour of the natural vegetation, some relationship between parent material and soil organic matter content is to be expected. One survey (Jones, 1973) has shown a small but significant difference in organic matter content between soils on basement complex and soils on sandstone in the Jebba-Kontagora area of Nigeria; and also, in a less typical area (West Cameroun), a larger difference between soils on basic rocks (basalts, lava etc.) and soils on acid rocks (basement complex).

(iii) *Effect of Topography*

Through soil erosion organic matter may be lost from sloping land and may accumulate at the foot of slopes. More important, however, is the effect of topography on soil drainage. Impeded drainage, whether topographical or pedological, causes a relative accumulation of organic matter, probably due to a longer growing season for the vegetation, a reduced microbial activity, and to less frequent and less severe flushes of mineralization at the start of the rains. Jones (1973) found that at all rainfall levels within the region poorly drained soils were considerably higher in organic matter than well-drained soils, although the difference tended to be less in soils of high clay content.

(iv) *Effect of human activities*

The degradative effect of cultivations on soil fertility in general (Bachelier *et al.*, 1956; Nye and Greenland, 1960; Watson and Goldsworthy, 1964) and on soil organic matter in particular (Maignien, 1948; Quantin, 1965) has frequently been remarked on.

There have, however, been few quantitative studies, for there is always the problem of obtaining accurate site histories. In the data of Gavaud (1968) from Niger, differences in organic matter status between virgin and cultivated sites were only small, but the duration of cultivation on the cultivated sites was not clear. In the more humid savanna of the Central African Republic, Quantin (1965) found that soil from an old fallow dominated by *Imperata cylindrica* and degraded by previous intensive hand cultivation had a carbon content about 20 per cent less than that of a normal savanna soil. In an area of the Nigerian Guinea savanna zone, soils under cultivation or under fallow following cultivation had a mean carbon content (0·58%) little more than half that of soils that appeared not to have been cultivated recently (1·03%), a difference that was statistically highly significant (Jones, 1973). Such a difference is considerably greater than the amplitude of the fluctuations about a mean soil carbon content under an alternation of cropping and fallowing (Greenland and Nye, 1959), presumably because the uncultivated sites more nearly represent a "bush control".

In temperate areas the decline in organic matter in the soil to new equilibrium levels is determined by environmental conditions and the farming system practised (Russell, 1973). Cultivation indices, i.e. organic matter in cultivated soils as percentages of that in virgin soils, of 40—73 per cent have been found for U.S.A. soils after 30—40 years cultivation (Haas *et al.*, 1957), and values as low as 30 per cent in some Indian soils (Jenny and Raychaudhuri, 1960). The Nigerian figures quoted above (Jones, 1973) yield a value of 56 per cent, but this was for a bush-fallowing system, whereas U.S.A. and Indian values were for continuous agriculture. With the increasing pressure on land in certain parts of the savanna, the area under continuous cultivation seems certain to increase and it is therefore instructive to look in some detail at two sites for which there are data of organic matter changes under continuous cultivation.

The changes in soil organic matter status at Sefa, Senegal since clearing from dense dry savanna woodland in 1950 have been described by Fauck (1956) after 6 years, and by Fauck *et al.* (1969) and Charreau and Fauck (1970) after 15 years of continuous mechanical cultivation. Their results can be criticized on the grounds that at each time of sampling the data for the cultivated soil were presented only in comparison with contemporary data from woodland control soils and not with previous data from the same cultivated plots. Changes occurring over the years in any one plot appear not to have been followed. Some precision has therefore been lost, but the changes were sufficiently great not to be obscured by it.

The main conclusions were:

(i) There was a large loss of organic matter as a result of clearing and cultivation. After 15 years cropping there had been losses of 18 and 40 tonnes organic matter per hectare respectively from the 0—20 cm horizons of a "sol rouge" (slightly Ferrallitic Soil) and a "sol beige" (leached Ferruginous Soil). However, comparison with results obtained nearly 10 years earlier suggested that most, if not all, of these losses had occurred during the first few years of cultivation and that, after this initial period was over, the soil organic matter status was maintained.

(ii) The loss of organic matter was largely from the top 10 cm (Table 22), and the cultivation indices for the top 15 cm were 52 and 32 per cent, respectively, for "sol rouge" and "sol beige".

Data from several long term experiments at Samaru, Nigeria (Jones, 1971) are summarized in Table 23. As with the Sefa results, it was not possible to trace the changes in organic matter content in one plot over the whole course of the experiment, and most of the comparisons are between different treatments sampled in the same year and between these and a mean value calculated for local soils under bush. However, it will be seen that where data exist for more than one sampling year (DNPK experiment, 1959, 1964, 1967), the result contrasts with the findings at Sefa. Even after 10 years cultivation there is no sign of any stabilization of organic matter level, and up to the eighteenth year, at least, losses were continuing at the annual rate of 4—5 per cent.

Cultivation indices after ten or more years of cultivation were 43 per cent or less. In the one experiment in which cultivation was not continuous (Gamba fallow rotation), but alternated with periods of planted grass (*Andropogon gayanus*) fallow, cultivation indices were higher in proportion to the length of fallow period.

Continuous cultivation over a period of 15—20 years reduced the organic matter of the topsoil of the Sefa and Samaru sites to between 25 and 45 per cent of what it is under natural vegetation. At Sefa, but not at

TABLE 22

Effect of 15 years cultivation on amounts of carbon (per cent) at different depths in two soils at **Sefa,** Senegal (Fauck *et al.*, 1969)

Depth, cm.	*"Sol rouge"*			*"Sol beige"*		
	Woodland Control	*Cultivated*	*Difference*	*Woodland Control*	*Cultivated*	*Difference*
0—5	1·42	0·58	—0·84	2·34	0·70	—1·64
5—10	0·96	0·51	—0·45	1·41	0·40	—1·01
10—15	0·69	0·51	—0·18	0·94	0·42	—0·52
15—20	0·52	0·51	—0·01	0·70	0·48	—0·22
20—25	0·43	0·49	+0·06	0·62	0·50	—0·12

TABLE 23

Effect of cultivation on amounts of organic carbon in topsoil (0-15 cm) at Samaru, Nigeria (Jones, 1971).

Description	*Remarks*	*Duration, years*	*% C*	*Cultivation index*
Local bush	Calculated mean		0·60	
Monocropping v rotation experiment	No manure or fertilizer	20	0·22	37%
DNPK experiment	No manure; some fertilizer. Sampled:			
	1959	10	0·26	43%
	1964	15	0·21	35%
	1967	18	0·17	28%
Gamba fallow rotation experiment	Fallow: cropping years ratio:			
	2:3	19	0·24	40%
	3:3	19	0·30	50%
	6:3	19	0·41	68%

Samaru, an equilibrium appeared to have been reached. Much depends upon the rate of return of crop residues to the soil, and the practice at Samaru of removing all above-ground crop residues from the field every year may well be the cause of the difference (Jones, 1971). This point is considered again below.

(b) *Turnover and decomposition rates*

It has been shown that cultivation usually brings about a decline in soil organic matter. Reasons for this include:

(i) A large loss of surface litter, standing vegetation, and perhaps some of the organic matter in the surface soil during the initial clearing and burning operation. Erosion losses may also be high at this time. Charreau and Fauck (1970) have commented on these effects during the initial clearing at Sefa.

(ii) Cleared and cultivated land is exposed to much greater insolation for a large part of the year than land under natural vegetation, leading to

(a) higher temperatures and more complete drying of the soil during the dry season, with a consequently greater flush of mineralization when the soil is subsequently rewetted (Birch, 1959; Enwezor, 1967);

(b) higher soil temperatures during the early part of the rains which, along with aggregate disruption and increased aeration resulting from recent cultivation, produce very high rates of oxidation of organic matter (Moureaux, 1967). This situation may continue well into the wet season, because even where crops are sown early, cover is established more slowly and is rarely as complete as under natural bush vegetation.

(iii) Cultivation, however carefully carried out, almost always leads to some increase in erosion, and any loss of topsoil involves a loss of organic matter.

Apart from the initial losses on clearing, these problems are perennial ones for both bush-fallowing and continuous agricultural systems; but, whereas bush-fallowing affords some restocking of the organic reserves at intervals, under continuous cultivation there are only the inputs from crop residues and, possibly, manure to offset the heavy losses that occur. It is the balance between these inputs and the rate of loss that determines the eventual equilibrium level of organic matter in a particular cropping system.

One aspect needs emphasis. From the agricultural point of view it is not only the absolute amount of organic matter in the soil that is important, but also its rate of turnover. For in the absence of fertilizers, crops are dependent upon the mineralization of organically-held nutrients for adequate nutrition, and, hence, a rapid breakdown of the organic matter during the growing season is essential. Where fallowing is practised, one of its main functions is to accumulate residues which will release nutrients rapidly to crops on recultivation. Rates of accumulation under fallow and subsequent decomposition under cropping in a bush-fallowing system have been discussed by Greenland and Nye (1959) and Nye and Greenland (1960).

The rate of change of soil carbon may be expressed by the equation:

$$I = A - kC$$

where I is the annual increase of carbon
A is the annual addition of carbon
C is the amount of carbon in the soil
k is the decomposition constant.

Although most authors refer to humus carbon, they appear not to distinguish between humus carbon and total organic carbon. The distinction is important since the rate of breakdown of the relatively large amounts of unhumified organic matter in the soil immediately after a fallow is much more rapid than that of the humus, and values of the decomposition constant, k, are correspondingly higher.

Data of Quantin (1965) from Bambari, Central African Republic, indicated annual losses of total carbon of 4·5—6·0 per cent during three years' cropping of land previously under *Imperata cylindrica* fallow; those of Fauck (1956) at Sefa a decline from 0·75 to 0·50 per cent carbon in 6 cropping years after clearing, or nearly 7 per cent per annum; and those for the Gamba fallow rotation experiment at Samaru (Jones, 1971) showed annual losses when the fallow was recultivated sometimes in excess of 10 per cent. These figures all refer to net losses, and since some residues, if only roots, are generally returned to the soil during cropping, the decomposition constants will have been somewhat higher.

Losses of humus carbon during the same period are likely to have been considerably less. Nye and Greenland (1960) estimated a decomposition constant for humus of 4 per cent in a cultivated savanna soil, which falls within the range 2—5 per cent suggested by Charreau and Fauck (1970) for Sefa. The Samaru data do not permit the calculation of a humus decomposition constant, but an annual loss of 4—5 per cent total carbon between the 10th and 18th years of the DNKP experiment on plots receiving no organic residues apart from crop roots suggests that it would fall within the same range.

Under bush fallow the annual addition of carbon (A) is difficult to quantify, and this may account for some of the divergence between estimates of humus decomposition constants. Greenland and Nye (1959) gave 0·5—1·2 per cent per annum for grass savanna and 2—5 per cent for tropical forest (both 0-30 cm); Charreau and Fauck gave 4—7 per cent for dense savanna woodland (depth unspecified). In cultivated soils "A" is more easily calculated. Weights of manure or compost added to the soil are usually known, and those of both the above and below-ground portions of the crop residues are either known or can be estimated.

When any organic material is added to the soil there is usually an initial period of rapid decomposition, during which part of the material is mineralized and part humified. Although obtained under temperate conditions, Monnier's (1965) values for this humification factor or "isohumic coefficient" are of interest in this context:

Cereal stover	0·08—0·20
Residues of grassland	0·15—0·45
Manure	0·15—0·45
Young green manure	about zero

They show that the annual addition of humus carbon is not determined solely by the mass of residues returned to the soil but varies widely according to the nature of the original material. Green manure, for instance, is rapidly decomposed in the soil and contributes little or nothing to the stock of humus.

Calculations from the data of the Samaru long-term experiments show that, relative to a control, about 12—15 per cent of the carbon and 24—31 per cent of the nitrogen applied as manure, and about 20 per cent of the carbon and 65 per cent of the nitrogen applied as groundnut-shell mulch appear to have been retained in the soil (Jones, 1971). No figures are available for crop residues. However, the retention in the soil humus of only 5—10 per cent of the 5—7 tonnes per hectare of stover dry matter from a good maize or sorghum crop would mean an addition of between 150 and 400 kg humic carbon per hectare, or approximately 0·01—0·02 per cent of the top 15 cm of soil.

Under stable conditions the amount of humus in the soil eventually achieves an equilibrium value when, since $I = 0$, $A = kC_0$; and the size of the equilibrium value (C_0) depends on the decomposition constant (k) and the annual addition (A). Charreau and Fauck (1970) have expressed the opinion that k is generally constant for a given pedoclimate and is therefore not amenable to much control by the agronomist. This seems improbable because the type, depth, frequency and timing of soil cultivations, and also the nature and rapidity of establishment of crop cover, through their effect on soil temperature and moisture conditions, can affect appreciably the rate of humus oxidation.

Soil fertility benefits from the presence of a high level of organic matter and also from a high rate of turnover of that organic matter. It will be seen from the preceding discussion that these two conditions are complementary, both being dependent upon a high rate of return of organic material to the soil. Under a bush-fallowing system of agriculture this requirement is fairly adequately met, particularly if the fallow can be protected from fire, but when cultivation is continuous it is important that provision be made for a substantial annual return to the soil of crop residues and, where available, manure and compost.

(c) *Effects on Soil Fertility*

Although crops can be grown in the absence of organic matter it is a common experience for them to be easier and cheaper to grow when it is present. As it has been shown above that organic matter contents of savanna soils are low and become even lower under cultivation, the effects of organic matter on soil fertility warrant careful consideration. We shall consider them under three general headings:

(i) nutrient supplies
(ii) physical effects
(iii) cation exchange capacity

(i) *Nutrient Supplies*

With fertilizer consumption still at a low level nearly the whole of the nitrogen uptake of non-leguminous crops is provided by the mineralization of soil organic matter during the season of growth. In general, the amount of nitrogen mineralized is approximately proportional to the total amount present so that, except when the soil C:N ratio is high, nitrogen availability increases with soil organic matter content. Assuming an annual mineralization of 4 per cent, a soil containing 300 ppm total N (0·03 per cent N) in the top 30 cm will, for instance, yield about 50 kg N/ha.

A relationship between soil C:N ratio and previous cropping history was demonstrated by Nye (1950; 1951) for savanna sites in Ghana, the ratio being highest at sites which had been left in fallow for long periods. As with total organic matter, climate is an important factor; the results of Gavaud (1968) indicate that C:N ratios tend to decline with decreasing annual rainfall, and some very low values are recorded for the soils of the most arid parts of the region (Maignien, 1959; Audry, 1961; Boulet, 1968). Jones (1973) obtained the relation:

$C:N = 7 \cdot 72 + 0 \cdot 0047 X$ ($r = 0 \cdot 476$***), where X = mean annual rainfall (mm).

From this one may calculate that the mean C:N ratio increases from 10·1 to 14·8 over the rainfall range, 500 to 1500 mm per annum. It is perhaps for this reason that the nitrogen deficiency found in the southern savanna, following clearing of tall grass (Nye, 1950; Greenwood, 1951), is less acute in the drier areas to the north (Goldsworthy, 1967a; b).

Organic matter may improve phosphate availability in two ways: first, through the mineralization of organic phosphorus, and secondly by humus adsorption on some soil mineral surfaces which would otherwise adsorb phosphate. Some authorities have suggested that the mineralization of organic phosphorus is more difficult than that of nitrogen; but when long-rested savanna sites in Ghana were cultivated, the C:N ratio decreased but the C:P ratio did not (Nye and Bertheux, 1957), implying more rapid mineralization of organic phosphorus than of organic nitrogen. If the mineralization rate of the organic P is put at 4 per cent, that is, if it is assumed to be the same as that for nitrogen, 30 ppm organic P in the top 30 cm of soil (Chapter 7) would release 5 kg P/ha, a significant contribution to the amount of phosphate available to crops.

Mineralization of organic sulphur components appears to parallel that of the nitrogen components. As about 80 per cent or more of the soil sulphur is present in the organic matter (Bromfield, 1972), its rate of mineralization may affect crop growth.

(ii) *Physical effects*

Infiltration, water-holding capacity, soil structure and resistance to erosion are among the soil physical properties that have been reported as being improved by increased levels of soil organic matter; and various experiments have shown that the general physical status of the soil improves under bush or grass fallow simultaneously with an increase in soil organic matter content (Martin, 1944; Pereira *et al.*, 1954; Morel and Quantin, 1964; Wilkinson, 1970). The improvement is, however, short-lived once cultivation is restarted (Pereira *et al.*, 1954; Stephens, 1960c; Combeau and Quantin, 1963), and both soil organic matter and, usually at a more rapid rate, soil physical properties deteriorate to their pre-fallow level. Nevertheless, the nature of the relationship between soil physical status and organic matter status has been a controversial issue. Nye and Greenland (1960) suggested that once "the ephemeral constitution built up in the surface layer under the fallow has been shattered by rain, the extent to which the deterioration will continue depends largely on the inherent constitution of the soil", and a particularly important factor in this inherent constitution is the bonding of individual soil particles into micro-aggregates by iron oxides.

Since that time a number of authors (e.g. Quantin, 1965; Fauck *et al.*, 1969; Charreau and Fauck, 1970) have continued to stress the importance of the soil structural state for satisfactory crop growth, and the dependence of this structural state on the maintenance of the soil organic matter status. Strong relationships have been demonstrated (Combeau and Monnier, 1961; Quantin, 1965) between soil organic matter content and the index of structural instability, I_s (Chapter 4). For the slightly Ferrallitic savanna soils of the Central African Republic, Combeau and Monnier (1961) obtained the highly significant relationship:

$$I_s = -1 \cdot 327\, C\% + 3 \cdot 1$$

That is, structural stability increased with increasing carbon content. Further work (Combeau and Quantin, 1964) indicated that it is the non-humified fraction of the soil organic matter that is responsible for maintaining soil structure. For a set of cultivated soil samples from Bambari, Central African Republic, it was found:

$$I_s = -2 \cdot 01\, C_{mo\text{-}mh} + 3 \cdot 16 \qquad (r = 0 \cdot 73)$$
$$I_s = +0 \cdot 133\, t_h - 1 \cdot 54 \qquad (r = 0 \cdot 81)$$

where $C_{mo\text{-}mh}$ represents total soil carbon less the phosphate-extractable humic fraction, and t_h the humic carbon as a fraction of soil carbon. Structural instability tended to increase as the proportion of the organic carbon present as fulvic acid increased, but there was no relationship with humic acid content. Similar, if less detailed, reports from the savanna areas of Congo Brazzaville (Niari valley) support the view that structural instability as measured by I_s is negatively correlated with the non-humified fraction of the soil organic matter (Martin, 1963; 1970).

The conclusion that it is the unhumified fraction of the soil organic matter that confers some stability on the structure accords with other experience. At Sefa, for instance, when savanna woodland soils were brought into cultivation, the physical state and workability of the soil declined while the degree of humification of the soil organic matter increased (Fauck *et al.*, 1969). In all soils taken out of bush or grass fallow the proportion of readily decomposable unhumified organic residues is high, and their rapid decline following the start of cultivation parallels the rapid decline in soil physical properties, such as rainfall infiltration (Pereira *et al.*, 1954; Wilkinson, 1970), a decline that is difficult to explain in terms of the relatively slow loss of humus. However, Wilkinson (1970) suggested that the effect of cultivations was to sever the channels made by soil animals, and it is probable they also sever root channels.

Part of the effect of the unhumified material may be simply mechanical. Russell (1971), in reviewing the role of organic matter in the stabilization of soil structure, noted that the undoubted, but usually short-lived, improvement of the structure of sandy soils following the use of farmyard manure or green manure may be due in part to partially decomposed plant debris mechanically preventing the sand particles falling down into close packing on wetting.

The dependence of soil structure upon the transient, unhumified organic matter fraction provides a justification for changing from bush-fallowing, as a method of soil restoration, towards a system of continuous cultivation with appropriate management of crop residues. A high level of unhumified organic matter in the soil can be maintained only by frequent additions of fresh residues. The advantage to be gained in terms of crop yield from ploughing in residues has been demonstrated at Sefa (Charreau and Fauck, 1970); and one practical approach at present being tried at Bambey is the breeding of cereals of sufficiently short growth period to facilitate early harvest and the incorporation of residues before the onset of the dry season. Such innovations, however, will not be for the hand farmer, since residue incorporation requires animal or mechanical traction. Zero tillage (direct drilling) may prove a more effective method of increasing the level of organic matter, but no experimental work in the savanna has yet been reported.

A further result of Combeau and Quantin (1964) merits mention. They found at Bambari that the soil water-holding capacity was related more closely to the amount of humified organic matter in the soil than to unhumified organic matter. If this result is more generally applicable, then in the more arid areas at least some thought will have to be given to the maintenance of humus *per se* and how it is affected by frequent additions of fresh organic material. The evidence from temperate countries, however, is that an increase in soil humus gives only a very small increase in the amount of water available to plants.

(iii) *Effect on cation exchange capacity*

The important contribution made by soil organic matter to the exchange capacity of savanna soils has frequently been pointed out (Nye and Greenland, 1960; Ollat and Combeau, 1960; Nye, 1963; Fauck, 1963; Hawkins and Brunt, 1965; Martin, 1970). The predominant clay mineral is usually kaolinite, and, as a result of eluviation, even this may be present in only small quantities in the topsoil. In a "sol beige" at Sefa the exchange capacity dropped from 6·6 to 3·75 me/100 g during the six years of cultivation that followed bush clearing as the direct result of the rapid decline in organic matter (Fauck, 1956 ; Fauck *et al.*, 1969).

Values for the exchange capacity of organic matter in savanna soils have usually been found to be about 350 me/100g carbon (de Endredy, 1954; Hawkins and Brunt, 1965; Jones, 1971). From the value of 368 me/100 g carbon found at Samaru, Nigeria, it was calculated that an increase in soil carbon content from 0·1 to 1·0 per cent would increase the exchange capacity from 0·98 to 4·28 me/100g soil and increase the contribution of the organic matter to the exchange capacity from 38 to 86 per cent (Jones, 1971). Any decline in the soil organic matter content, therefore, involves a significant loss of exchangeable bases. Nye and Greenland (1960) calculated that the increase in organic carbon in a savanna soil during a ten year fallow might be about 1350 kg/ha, which would increase the calcium storage of the topsoil, through increased exchange capacity, by about 100 kg/ha. In the Sefa experiment mentioned above (Fauck, 1956) exchangeable calcium declined by as much as 1 me/100g soil as a result of six years cultivation.

2 CATION EXCHANGE CAPACITIES

As will be seen from Tables 8 to 13, cation exchange capacities of most soils are low, often being less than 5 me/100 g. The exceptions are Ferrisols, Eutrophic Brown Soils and Vertisols which occupy only a small proportion of the land area.

It was mentioned above that the organic matter contents of soils are important in determining the actual exchange capacities. This is because the clay contents are often low (below 15 per cent) and the clay is mainly kaolinitic, so that the contribution of the clay to the cation exchange capacity is normally no greater than 1—2 me/100g soil and is frequently less. In Ferrisols, Eutrophic Brown Soils and Vertisols clay contents are higher and 2:1 lattice silicates are present. Higher clay contents also occur as textural B horizons in some Ferruginous Soils (sols ferrugineux tropicaux lessives) and give rise to somewhat higher exchange capacities.

Although the cation exchange capacities quoted above are low, they overstate the capacity of the soil to retain cations under field conditions. This is because the charge characteristics of the exchange surfaces (humus and kaolinite) are pH-dependent and many soils have a pH value in water of 6·0—6·5 (Table 7) or less, whereas exchange capacities are usually measured at pH 7·0 with molar ammonium acetate. The ammonium acetate method may overestimate the cation retaining properties of the soils under field conditions for another reason, for Barber and Rowell (1972) found with an iron-rich kaolinitic soil from Tanzania that the negative charge and cation exchange capacity were much less in low electrolyte concentration than in 1M concentration.

The low exchange capacities have the following implications:

(i) Many soils are low in the exchangeable cations required as crop nutrients, and under cropping systems where nutrient demand is high they will quickly become unable to meet these requirements.

(ii) A low content of exchangeable cations may lead to nutrient imbalance from conventional fertilizer practices, in which only one or two nutrients are applied. It may therefore become necessary to balance the nutrient supply by using more complete fertilizers.

(iii) Nutrients applied as cations, ammonium, potassium, magnesium etc., are weakly held and leaching may occur from the surface soil. Calculation shows that a soil with a cation exchange capacity of 1 me/100g would be saturated to a depth of 15 cm by the ammonium ions given as an application of 280 kg N/ha, and although these are extreme conditions they are not entirely unrealistic.

(iv) Rapid soil acidification occurs where ammonium sulphate and, to a lesser extent, urea are used. Other changes from traditional practice, for example the abandonment of burning or of the fallow break, may also accelerate acidification.

These implications are general for soils throughout the world but they are obviously more important where cation exchange capacities are low. In the savanna they will assume increasing importance as nutrient demand intensifies with the advent of improved crop varieties and permanent agriculture.

3 SOIL pH AND BASE SATURATION

Under natural vegetation and traditional farming systems it is unusual to find soils more than slightly acid. The mean pH values quoted in Table 7 for surface samples of Brown and Reddish Brown Soils of Arid and Semi-Arid Regions, Ferruginous Soils and Ferrallitic Soils are respectively 6·8, 6·2 and 6·0 (measured in water, usually at soil:water ratios of 1:1 to 1:5). Values for individual soils are rarely below 5·0, although some Hydromorphic Soils are more acid, and Beye (1972) has reported pH values as low as 3·0—4·5 on acid sulphate soils in the Senegal river delta.

At first sight one might expect the soils to be more acid than they are, for they are subjected annually to through-drainage. Factors that probably restrict the development of acidity are (a) the return of bases to the surface by the "pumping" effect of vegetation, (b) the effect of burning, (c) the restriction on the leaching of bases imposed by the low anion content of the soil solution, (d) the low content of soil humus, (e) the low net rate of removal of bases by cropping because of low yields and the return to the soil of much of the crop residue, (f) the very small and non-intensive use of ammonium fertilizer and urea, and (g) the absence of industrial pollution by acidic gases. Some of these factors were discussed by Nye and Greenland (1960) and will not be discussed fully here, but one or two comments are necessary.

First, pH values, exchangeable bases (Ca, Mg, K, Na), and per cent base saturation tend to be higher in the surface horizons than at depth in profiles of Ferruginous and Ferrallitic Soils (Tables 8, 9). The difference is probably due to the return of cations to the soil surface through the pumping effect of the vegetation noted above, although pH changes with depth in other soil Classes are less clear. However, this process operates only under natural vegetation, for most crops are relatively shallow rooted, and surface soils tend to become more acid under continuous cultivation, whether acidifying fertilizers are used or not (Fauck *et al.*, 1969; Charreau and Fauck,

1970). The need to neutralize acidity by applying lime is discussed in Chapter 10.

Secondly, both vegetative growth and leaching intensify southwards across the savanna, so that the pumping effect is more important for the maintenance of high pH and base saturation in the more humid areas of the south. However, the relationship with rainfall is not necessarily a simple one. Scott (1962) in East Africa related base saturation at 75—100 cm depth (to avoid erosion and environmental effects at the surface) to mean annual rainfall. In arid areas base saturation decreased exponentially with increasing rainfall, but above about 750 mm rainfall the effect of vegetation asserted itself and base saturation increased again to a maximum at about 1150 mm rainfall. As rainfall increased still further, leaching again became dominant and base saturation declined. No similar survey of these trends appears to have been made in West Africa.

The final point concerns the unfortunate choice of 1M ammonium acetate at pH 7 for measuring cation exchange capacities and hence per cent base saturation. Because the soils have usually a pH lower than 7, and a high proportion of their electrical charges are pH dependent, this method will, as discussed earlier, overestimate the ability of the soils to adsorb cations under field conditions. It would be better either to measure the cation exchange capacity at the electrolyte concentration and pH of the soil, or to sum all exchangeable cations including H, Al and Mn, or to measure electric charges over a range of pH values.

Chapter 6

NITROGEN

Savanna soils are generally low in total nitrogen. Among the data for 295 well-drained topsoils from all over the region, examined by Jones (1973), values ranged from 0·008 to 0·290 per cent, with a mean of 0·051 per cent. These low values are closely linked with the generally low levels of organic matter in these soils; the mean C:N ratio for these 295 sites was 11·7 and the correlation between total nitrogen and organic carbon was high ($r = 0{\cdot}937^{**}$). As with carbon, more than half the observed variability in soil nitrogen content could be accounted for in terms of a multiple linear regression on soil clay content and mean annual rainfall.

The emphasis in this chapter is on the nitrogen cycle in the absence of fertilizers. We consider first the losses and gains of total soil nitrogen under bush fallows and under cultivation and, secondly, the factors affecting nitrogen availability to crops. Nitrogen fertilizers are discussed only to the extent that they are involved in soil processes, and discussion of their effects on crop yields is left to Chapter 10.

1 GAINS OF NITROGEN

The nitrogen content of a soil is determined by the relative rates of gain and loss. Losses usually predominate under cultivation, and gains under bush fallow, but the situation is a dynamic one, and under stable conditions nitrogen content tends towards an equilibrium value. In general the rate of build-up under fallow depends on how far the soil nitrogen content is removed from equilibrium: the further removed, the greater the rate of build-up.

However, if the soil is too degraded chemically or physically, the establishment of the natural vegetation is retarded (Morel and Quantin, 1964), and nitrogen build-up is correspondingly slower. The effect of soil nutrient status on nitrogen build-up was seen in planted Gamba grass (*Andropogon gayanus*) fallows at Samaru, where mean annual increases in topsoil nitrogen were 83 kg/ha in plots that had received phosphate fertilizer some years previously but only 59 kg/ha in plots that had not. These values appear high compared with the estimate by Nye and Greenland (1960) of 10 kg nitrogen/ha annual gain under high grass savanna, but their value was based on a soil organic matter status 75 per cent that of undisturbed natural vegetation and took account of an annual loss through bush burning of another 28 kg/ha. The Samaru plots were at a lower organic matter status relative to that of local soils under bush and also were protected from fire.

In the absence of nitrogen fertilizers all the nitrogen in the soil-plant system originates in atmospheric deposition (mainly rainfall) and biological fixation.

(a) *Rainfall*

The limited evidence available suggests that over much of the savanna the annual contribution, as mineral-N, is not very large. Thornton (1965), in Gambia, found a relatively high deposition of 47·1 kg/ha at a station 15 km from the sea, but this tailed off to 40·1 and 12·7 kg/ha/annum, respectively, at stations 130 and 218 km inland; and at Samaru, Nigeria, more than 800 km from the sea, the amount is 4—5 kg/ha/annum (Jones and Bromfield, 1970). Moreover, it has been suggested in respect of other non-maritime tropical areas (Meyer and Pampfer, 1959; Wetselaar and Hutton, 1963) that much of the mineral-N in the rain derives from the soil and is not a true accession.

(b) *Symbiotic fixation*

Biological fixation may be either symbiotic or non-symbiotic, their relative importance often being uncertain. Nye and Greenland (1960) estimated a net annual fixation in the soil-plant system of the tall-grass savanna of about 38 kg/ha/annum. However, although leguminous trees and shrubs are frequent in the savanna, these authors doubted whether they support many active nitrogen-fixing bacteria, for Nye (1958a, b) had shown that in soil very low in nitrate the nitrogen content of leguminous species was generally low.

Nevertheless, it should not be assumed that trees and shrubs make no contribution to the soil nitrogen economy anywhere in the savanna. Dommergues (1966) reported fixation in Senegal of 60 kgN/ha by stands of *Casuarina*, a non-leguminous species, although since *Casuarina* is not native to West Africa this can be of only local importance.

Of greater agricultural value is *Acacia albida*, a traditionally protected tree over much of the drier savanna. This tree is leafless during the rainy season, and crops grown underneath it have often been observed to give markedly higher yields than those grown outside its influence. Studies have shown the soil under these trees to be generally richer, particularly in organic matter and nitrogen, exchangeable calcium and total exchangeable bases, than that of the surrounding area (Charreau and Vidal, 1965; Radwanski and Wickens, 1967). In the Sudan, Radwanski and Wickens (1967) noted that the C:N ratios of soils under *Acacia albida* were unusually low, and from Senegal Dancette and Poulain (1969) reported the results of sorghum trails which suggested that an additional 15—20 kg N/ha was mineralized in them. Such results suggest vigorous nitrogen fixation, but Radwanski and Wickens (1967) reported that few nodules were seen in profile pits. It may be that any fixation that occurs is non-symbiotic, encouraged by the improved soil mineral and organic status built up by the trees.

Although legumes have often been tried experimentally as fallows and in pastures in West Africa (e.g. Vine, 1953; Singh, 1961; Moore, 1962), there appear to be no published reports from the savanna region of the effects of these on soil and crop nitrogen. Vine's work related to results at Ibadan, Nigeria, in the forest/savanna mosaic. Later in a pot trial with *Centrosema pubescens* at the same station, Moore (1963a) reported that the soil-plant system gained 112—224 kg N/ha but, since in two of the soils large gains (112—148 kg N/ha) were also recorded under grass (*Eleusine coracana*), it was not clear how much fixation was due to symbiotic organisms.

Of the leguminous crops grown in the savanna the grain legumes, e.g. groundnuts and cowpeas, occupy the greatest area. Elsewhere there have been many studies on the nitrogen-fixing efficiency of crop legumes (Nutman, 1965; Vincent, 1965), but in West Africa nearly all the work has been of an agronomic nature and has provided no information about the amounts of nitrogen fixed nor how much is left in the soil at the end of the season. The haulm of grain legumes makes very good fodder and is rarely returned to the soil, so that any useful nitrogen build-up must come from the roots and nodules left in the soil. Fauck (1956) reported that a crop of groundnuts increased soil nitrogen at Sefa by more than 250 kg N/ha, but this is an exceptionally high figure and must be viewed with suspicion. In the Niari valley, Congo (Brazzaville), Martin (1970) found a decline in total nitrogen in a soil cropped to groundnuts for three years, relative both to the original savanna soil and to other plots bare-fallowed for three years. The assumption often made that the grain legume phase of a cropping cycle neither increases nor depletes the total soil nitrogen may well be a fair one, although some increase in levels of available nitrogen to the subsequent crop may be expected, as discussed below.

(c) *Non-symbiotic fixation*

Almost all the older accounts of non-symbiotic fixation in savanna soils were of a qualitative rather than quantitative nature, reporting where non-symbiotic fixing organisms were to be found, but giving no measure of their activity nor even always of their numbers.

In Senegal, Dommergues (1956b) found low numbers of both *Azotobacter* and *Beijerinckia spp* in cultivated and uncultivated soils in Casamance and Boulel-Kaffrine, but not in cultivated soils of the Cape Verde peninsular (Dommergues, 1959; 1960). Moore (1963b), in Nigeria, found *Azotobacter* in most of his upland savanna soil samples, *Beijerinckia* occurred more rarely, but *Clostridium spp* were always present. This is in agreement with the experience of Meiklejohn in both Rhodesia (1954; 1968b) and Ghana (1962), where *Clostridium spp* were more numerous than either *Azotobacter* or *Beijerinckia* in both forest and grassland soils. However, population counts of *Clostridia* are notoriously poor indicators of their activity (Jensen, 1965), so that their higher population and more frequent occurrence cannot be taken as evidence of any greater importance in nitrogen fixation. Unfortunately, none of the counts that have been made gives any real indication of the significance of the organisms in the nitrogen economy of the sampled soil.

Odu studied amounts of nitrogen fixed in cultivated and uncultivated soils from the derived savanna of Western Nigeria (Odu, 1967; Odu and Vine, 1968a). In an incubation experiment involving six cycles of wetting and drying, total nitrogen increased by approximately 20 ppm, but in an N^{15} study in which the soil was incubated for four weeks at field capacity, fixation was only 0·9—1·6 ppm. It was suggested that wetting and drying assisted fixation by releasing decomposable organic material to the nitrogen fixing bacteria. Another finding was that the amount of nitrogen fixed was little affected by the cultivation history of the site or by the related soil organic matter content.

From this very limited information no reliable estimate can be given of the significance of non-symbiotic fixation in savanna soils. Fixation may be regarded as a response to a low level of nitrogen availability in the environment, and this condition is widespread. Organisms capable of fixing nitrogen are usually present, but low availability of phosphate, molybdenum and energy sources may frequently limit their effectiveness.

The above reports referred to non-symbiotic fixation in the bulk of the soil. However, more recent work, stemming initially from studies on sugar-cane and tropical grasses in Brazil and employing the acetylene-reduction technique to assess nitrogenase activity, has indicated that fixation by free-living organisms (e.g. *Azotobacter paspali*) in the rhizosphere of non-leguminous plants may be of greater importance to the nitrogen economy of tropical soils (Kass *et al.*, 1971; Döbereiner *et al.*, 1972a; 1972b).

Studies on material from West Africa have shown that rhizosphere fixation activity varies widely with both soil type and plant species (Weinhard *et al.*, 1971). High values have been obtained under rice, rather lower values under *Eleusine coracana* and other grass species. It seems probable that maximum fixation occurs in ecosystems with the following features: wet soil, sunny climate and root systems that produce large amounts of exudates (Dommergues *et al.*, 1973). With the exception of rice, all plants so far found that stimulate such activity possess the C—4 dicarboxylic acid photosynthetic pathway rather than the C—3 Calvin cycle of most temperate plants (Dart and Day, 1975). The only field data available from the savanna are those of Balandreau and Villemin (1973), who measured a daily fixation of 25g nitrogen/ha in grassland in Ivory Coast.

These findings suggest an explanation for the results of Moore (1963a) at Ibadan, referred to above, in which the soil-plant system under *Eleusine coracana* gained the equivalent of 112—148 kg N/ha. They also provide a plausible hypothesis to account for the continued uptake of nitrogen by millet plants in Senegal soils when little or no mineral nitrogen could be found in the bulk of the soil (Blondel, 1971c, d, e, f).

It is therefore probable that non-symbiotic fixation in the rhizosphere makes some contribution to the nitrogen economy of the soil-plant system in fallows and under arable crops, but the extent and importance of the phenomenon are still to be assessed.

2 LOSSES OF NITROGEN

(a) *Burning*

The main agent of nitrogen loss from natural savanna is fire. Vidal and Fauche (1962), in lysimeter studies at Bambey, reported a loss of 20 kg N/ha/annum as a result of fallow burning, which agrees well with the earlier estimate of Nye and Greenland (1960) of about 28 kg N/ha/annum for the denser vegetation of the high-grass savanna.

(b) *Cultivations*

The rough balance existing between losses through burning and gains from biological fixation and rainfall is upset by the onset of cultivation. Even if there are no losses through erosion, increased soil temperatures and increased aeration enhance mineralization; there is an increased risk of loss through leaching and denitrification; and nitrogen is exported from the field in crop yield and lost to the air when residues are burned. Moreover, the elimination of native legumes and the rhizosphere population of grasses, and the decreased return of plant residues to the soil, reduce the potential for biological fixation. The rate of decline in soil total nitrogen in the first years of cropping may be reduced a little by the spreading of manure, the cultivation of leguminous crops, and even by fertilizer application, but the evidence strongly suggests that some decline is almost inevitable.

Published data for nitrogen losses from savanna soils on cultivation are collected in Table 24, all, unfortunately, from higher rainfall areas and most from mechanically cultivated sites. It can be seen that losses may be as high as 200 kg N/ha/annum. At Sefa, losses were of this order during the first few years after clearing, but later appeared to stabilize, and the subsequent decline was small (Fauck, 1956; Fauck *et al.*, 1969). Sefa lies within the zone of dense dry savanna woodland and its soils appear to exhibit the forest zone trait of having a concentration of organic matter in a humic surface horizon. The much greater relative decline in nitrogen content at the surface (Table 24) may reflect not only a more rapid mineralization in this horizon, but also erosion losses and mixing with soil from deeper horizons during cultivation (Fauck *et al.* 1969).

At Samaru mean annual losses of 25 kg N/ha were recorded even after more than 10 years of continuous cultivation, representing an annual decline in total-N of more than 4 per cent (Jones, 1971). Percentage losses were much higher during the three years of cultivation that followed three and six years under grass fallow, suggesting that the fallow was very effective in accumulating readily mineralizable reserves.

3 AVAILABLE NITROGEN

Although the amount of mineral-N (that is, ammonium-N + nitrate-N) available for plant growth is primarily dependent upon the rate at which soil organic nitrogen is mineralized, other soil processes tend to reduce it, and the growing plant must compete for its nitrogen nutrition against the effects of immobilization, denitrification and leaching. Actual uptake therefore is not determined solely by the nature of the soil but also by the depth and proliferation of the rooting system.

(a) *Soil and vegetation effects*

Since soil mineral-N comes from the decomposition of soil organic matter, there is a widely held expectation that availability of nitrogen will be proportional to organic matter status, and in general terms this is true (Scarsbrook, 1965). Such a relationship has been shown to exist in many different parts of the world, particularly in the U.S.A. (Allison and Sterling, 1949; Munson and Stanford, 1955; Klemmedson and Jenny, 1966), but also in Kenya (Robinson, 1968), southern Nigeria (Moore, 1960) and in a wide range of African soils, including many from the savanna (Gadet, 1967). However, few authors claim the relationship to be a very close one, and for any one soil total nitrogen content is a poor guide to probable nitrogen availability. Other factors may have a predominant influence.

Soil texture is one such factor. Saunder and Grant (1962), in Rhodesia, reported that rates of nitrogen mineralization both in incubation experiments

TABLE 24

Nitrogen losses under cultivation

Site	Mean annual rain, mm.	Notes	Sampling depth, cm	Years since clearing	% N in soil		Annual loss of N		Ref.
					Initial or Control	After cultivation	% of original	kg/ha	
1. Sefa, Casamance	1350	1st 6 yrs. after clearing savanna woodland	0—15	6	0·15	0·10	5·6	187	a
		12 years after clearing	0—4	12	0·085	0·030	5·4	28	b
			4—8	12	0·053	0·035	2·8	9	b
			8—12	12	0·036	0·027	2·1	5	b
			12—16	12	0·031	0·028	0·8	2	b
			16—20	12	0·027	0·031	—	—	b
2. Niari, Congo Rep.	1300	3 yrs. g'nuts following clearing	0—15	3	0·173	0·141	6·2	239	c
3. Grimari, Central African Republic	1500	Mech. cult.							
		+ fertilizer	0—15	4	0·093	0·093	0·0	0	d
		— fertilizer		4	0·093	0·082	3·0	62	d
		Hand cultivation		4	0·093	0·079	3·8	78	d
4. Northern Ghana	1000		0—30	6	0·055	0·046	2·7	96	e
5. Samaru, Nigeria	1100	11th—18th years of cultivation	0—15	—	0·026	0·017	4·3	25	f
		After:							
		6 years grass fallow	0—15	3	0·046	0·034	8·7	99	f
		3 years grass fallow	0—15	3	0·036	0·026	9·3	75	f
		2 years grass fallow	0—15	3	0·027	0·026	1·2	7	f

References: a, Fauck (1956); b, Fauck *et al.* (1969); c. Martin (1970); d, Quantin (1965); e, Stephens (1959); f, Jones (1971).

and in the field depended mainly upon the soil clay content. In the field an annual release of 2—3 per cent of the total-N in a clay soil increased to 3—4 per cent in a sandy-loam and to 4—5 per cent in a sand or loamy sand. While there are no data of this nature from West Africa, the high or very high coefficients of nitrogen mineralization recorded by Dommergues (1959) for a range of Senegal soils were probably a reflection of the generally sandy nature of these soils.

A second factor is soil C:N ratio. In a series of fertilizer trials in different parts of Ghana, Nye (1950; 1951) found an acute shortage of available nitrogen in land which had been left under fallow for ten years or more; and many crops responded well to applied nitrogen. On land more frequently cropped responses were smaller, implying a greater availability of native mineral-N. Similar results were obtained by Stephens (1960c) and Singh (1961) also in Ghana, by Greenwood (1951) in Nigeria, and much earlier by Rounce and Milne (1936) in East Africa. Moreover, the general practice among farmers in the savanna of avoiding nitrogen-demanding crops in the first year after a long fallow appears to reflect a similar experience.

Nye related his results to the soil C:N ratio, which was lower (11·5—13·9) after short fallows than after long ones (13·3—17·4), and lower still on continuously cultivated land (8·2—11·9).

The vegetation itself has an effect. Low levels of mineral nitrogen, particularly of nitrate-N, under grassland have been reported from many parts of the world, tropical (Wetselaar and Norman, 1960), sub-tropical (McGarity, 1959; Botha, 1963) and temperate (Richardson, 1938). The severity of the effect would seem to depend, at least in part, on the age of the grassland (Richardson, 1938; Theron, 1951; Greenland, 1958). The duration of low nitrogen availability upon subsequent reploughing has been variously reported as being either for one cropping season (Rounce and Milne, 1936; ap Griffith, 1951) or for up to three years (Nye, 1951) and to vary according to grass species (Mills, 1953). Depressed mineral-N levels have also been noticed under arable crops (Clark, 1949), and in this case the depression was related to the total weight of roots in the soil. When the grass or other crop is growing, the low content of mineral-N in the soil may be explained by rapid uptake by the roots and microbial immobilization, and the latter may continue after ploughing. These effects, which are not restricted to the savanna but are world-wide, have been reviewed by Harmsen and Kolenbrander (1965).

(b) *Climatic relationships*

The alternation of wet and dry seasons enforces a cyclic pattern of activity on the microbial population of the soil and on the availability of mineral-N to crops. Several aspects of this climatic effect may be distinguished.

(i) *Temperature*

High insolation and, in the more arid areas and on fields cleared for cultivation, a partial or complete lack of vegetative cover bring soil temperatures to a peak towards the end of the dry season. According to Moureaux (1967) temperatures of the surface soil may exceed 50°C. Soil biological activity is severely limited at this time by the lack of moisture, but with the onset of the rains, which improve soil moisture status without immediately causing any marked drop in soil temperature, there is an explosion of microbial activity. Moureaux (1967) commented on the high level of biological activity at temperatures between 55° and 65°C. He reported that maximum ammonification occurred at about 50°C in the more arid soils of northern Senegal but at about 25°C in a slightly Ferrallitic Soil in the south. A similar, though less pronounced adaptation to the local conditions was also noticed for nitrification, but in all soils this activity declined markedly above 40°C.

(ii) *Moisture*

It has been observed that nitrification is inhibited more easily than ammonification by high or low soil moisture content (Robinson, 1957; Moureaux, 1967). In the dry season in northern Nigeria Wild (1972a) found that ammonification continued slowly at water contents below 15 bars but nitrification was arrested. By the start of the rains there was a high ratio of NH_4-N: NO_3-N in the soil, which was reversed when the soil became sufficiently wet to permit nitrification.

Greenland (1958) suggested that nitrate-N might accumulate at the end of the rains when the soil is still moist, and there is some evidence for such an accumulation following groundnuts harvested soon after the end of the rains (Wild, 1972a).

(iii) *Effect of wetting a dry soil*

Closely related to both temperature and moisture factors is the flush of mineralization that occurs when a soil is remoistened after a period of dryness or heating. The magnitude of the flush depends on both the intensity and duration of heating and drying, so that, in general, the longer and hotter the dry season the greater the proportion of the soil nitrogen mineralized when the rains start again. The most detailed investigation of the phenomenon was the series of laboratory studies of Birch (1958; 1959; 1960), who also discussed the possible mechanisms involved. Results of field studies have been given by Hagenzieker (1957) and Semb and Robinson (1969) in East Africa and by Greenland (1958) and Blondel (1967; 1971c, d, e) in West Africa.

Birch (1958) showed that repeated cycles of soil wetting and drying led to repeated flushes of mineralization at each rewetting, the magnitudes of which declined only slowly with repetition. He pointed out that where the start of the rains is marked by a number of well-spaced storms interspersed with periods of intense soil drying, repeated flushes of mineralization might lead to considerable accumulations of nitrate. There appear to be no reports of this from the field, but Greenland (1958) showed that where the rainfall has a two-peak pattern a second flush may occur after the mid-season dry period.

The amount of mineral-N produced in the flush may be of considerable agronomic importance. Blondel (1971c, d) reported 58 kg N/ha mineralized in the top 100 cm of soil at Bambey (four-year mean) and 157 kg N/ha at Sefa (two-year mean); and Semb and Robinson (1969) obtained a range of 13 to 183 kg N/ha for topsoils at 13 sites in East Africa. At Samaru, Wild (1972a) found that the amount of mineral nitrogen in the topsoil at planting time depended on the previous crop and on the previous use of farmyard manure. With no manure the amount averaged 7 kg N/ha after cotton, sorghum or groundnuts but was as high as 70 kg N after groundnuts in soils that had received 12·5 t/ha of manure annually for 20 years. Thus in some circumstances, if uptake is efficient, enough mineral-N may be produced to support a moderate cereal crop.

Many authors (e.g. Birch and Emechebe, 1966) have attributed the widely attested benefit gained from planting crops early to efficient use of this early abundance of available-N, before it is lost by leaching. Undoubtedly this is so in some areas, particularly if the soil is very sandy and the onset of the rains very sudden. For instance, at Bambey in 1966 (Blondel, 1971c) nitrate-N reached a maximum of 35 kg/ha in the top 20 cm of soil fifteen days after planting, but this figure was halved during the next five days by 20 mm rain, and over the next two or three weeks nitrate-N was completely eliminated from the profile (0—100 cm) by further heavy rain (68 mm). Even though millet was planted immediately after the first rain, the whole of the flush of mineral-N, apart from some ammonium-N in the subsoil, had disappeared from the rooting-zone before the period of maximum plant demand had started.

However, this is not typical of all the savanna. Work at Samaru (Wild, 1972b) has shown very much slower leaching rates than those at Bambey; more native mineral-N appeared to be available to late-planted crops than to early ones (Jones and Stockinger, 1972), and decreases in yield occurred with late-planted crops even when adequate fertilizer nitrogen was supplied. In East Africa, too, yields of late-planted crops were not increased by fertilizer nitrogen (Goldson, 1963; Evans, 1963; Akehurst and Sreedharan, 1965; Turner, 1966).

The increase in mineral-N on wetting a dry soil is not universal. Simpson (1960) in Uganda and Saunder and Grant (1962) in Rhodesia found little evidence of any flush. Instead, Simpson reported an accumulation of nitrate during the dry period and postulated that the sudden entry of rain into this warm dry topsoil, rich in nitrate, led to conditions under which nitrate was easily lost by leaching and denitrification. Any flush of mineralization would probably have been masked.

(c) *Nitrification*

The term nitrification is applied here solely to the microbial oxidation of ammonium nitrogen through nitrite to nitrate. (Except where otherwise stated, nitrite is assumed to be present only in negligible amounts). This restricted definition thus excludes heterotrophic transformations of organic nitrogen into nitrates (Alexander, 1965), which, apart from one report (Odu and Adeoye, 1970), appear not to have been studied in the savanna. Restricting the definition to autotrophic transformations makes it possible to distinguish between mineralization (i.e. ammonification) and nitrification (i.e. nitrate formation from ammonium), and in view of the many reports there have been of slow nitrification in savanna soils (Nye and Greenland, 1960; Ahn, 1970), the distinction is an important one.

The phenomenon of slow nitrification has been attributed both to low numbers of autotrophic nitrifying bacteria and to the repressive effects of grass roots on soil mineral-N levels in general (see above). There is little doubt that the numbers of nitrifiers are often low in these soils (Meiklejohn, 1962; Odu, 1967; Wild, 1972a), although the results of Dommergues (1956b) suggest that they may be higher in areas of savanna woodland where grass is less dominant, and Wild (1972a) found much higher populations in soils under continuous arable cultivation than in those under grass fallow. The presence of grass appears to be significant.

One reason put forward to account for these low numbers has been the lack of ammonium substrate due to uptake by grass roots and heterotrophic organisms (Robinson, 1963; Odu, 1967; Wild, 1972a). This view is supported, circumstantially, by reports of increased populations at the beginning of the rains (Visser, 1966; Meiklejohn, 1968a), when there is usually a flush of ammonification, and after the addition of ammonium fertilizers (Dommergues, 1956a; Robinson, 1963; Meiklejohn, 1967; 1968a). Other nutritional factors may sometimes be important (Chase *et al.* 1968); for instance, at Bambey liming increased both the nitrifying population and the rate of nitrification, and at Sefa nitrifying activity was stimulated by lime and phosphate (Dommergues, 1956a).

Slow nitrification has also been attributed to the production by the roots of certain grasses of substances that inhibit nitrifying bacteria (Theron, 1951; Greenland, 1958). Over the years evidence has accumulated to support this hypothesis. Specific compounds, produced by the roots of grasses (Stiven, 1952; Hesse, 1956; Boughey *et al.*, 1964; Munro, 1966) and by other plants (Molina and Rovira, 1964; Rice, 1964; 1965), have been found which are capable of inhibiting nitrification, but their importance in the nitrogen economy of savanna soils has never been assessed.

Nitrification involves the conversion of a slowly leached cationic form of nitrogen to an anionic form, susceptible to both leaching and denitrification. It follows that the inhibition of nitrification may confer an agronomic advantage by conserving mineral-N in a form less easily lost from the rooting zone. However, it is doubtful whether this inhibition lasts long once the land is put under cultivation. As mentioned already, Wild (1972a) found much higher numbers of nitrifiers in arable soils than in grassland soils; and further work at Samaru (Table 25) suggests that after six years of grass

TABLE 25

Rates of nitrification of added ammonium-N in incubated Samaru soils (Jones, unpublished).

	Soil history					
	Years under grass fallow			*Years since the end of six years grass fallow*		
	2	4	6	1	2	3
Rate of nitrate production during first 10 days incubation, ppm/day.	1·25	0·80	0·69	2·97	2·09	2·86

fallow one year of cropping is sufficient to bring the nitrifying power of the soil back to the usual arable level.

A rather different situation has been reported from Bambey (Blondel, 1971a, c). Although ammonium sources added to a cropped soil at the start of the season were always rapidly nitrified, side-dressings applied later appeared not to nitrify at all. Ammonium levels declined quite rapidly, and this was attributed to plant uptake and microbial immobilization, but no nitrate was found. It was suggested that this was due either to the disappearance of the "microsites" favourable to the activity of the nitrifiers or to the vigorous competition from heterotrophic microflora.

(d) *Immobilization*

Reference has already been made to the immobilization of mineral-N by microbial assimilation. The various aspects of this process have been reviewed by Bartholomew (1965). Essentially, a certain quantity of nitrogen is assimilated into microbial tissue during the decomposition of plant residues in soil, and its subsequent release as mineral-N is usually slow. Where the residues are relatively rich in nitrogen, decomposition may give a net release (mineralization), but where the residues are poor, nitrogen is taken from the environment, in particular from the soil mineral-N.

It is this depletion of soil mineral-N that is generally referred to as immobilization. In a moist soil, in which some breakdown of organic matter is always going on, the two opposing processes of mineralization and immobilization proceed simultaneously. What is important in determining nitrogen availability is the balance between them, and this is controlled by the amount, nature and frequency of residues returned to the soil.

In the savanna, residues from both the natural vegetation (Nye, 1958a, 1959) and most crops are poor in nitrogen and seem likely to initiate a period of immobilization when returned to the soil. Nye's earlier results (1950; 1951) showed that nitrogen deficiency was greatest at sites recently cleared from long fallow, having high C:N ratios, the implication being that the presence of undecomposed plant residues in the soil encouraged vigorous immobilization. Soils of the grass-dominated savannas, before and for a time after clearing, are therefore seen as systems in which mineral-N is reabsorbed as fast as it is mineralized, and this situation continues until the carbonaceous sources of energy for the soil population are depleted.

This view must, however, be seen in the light of three studies involving immobilization in savanna soils. In the first, Greenland and Nye (1960) applied heavy dressings of straw to the soil, either by incorporation or as a mulch, but found that very little immobilization occurred. As the C:N ratio of the straw was about 70, this result was rather surprising. A possibility was that non-symbiotic nitrogen fixation occurred in the straw, and there was some evidence of this on mulched plots. Where straw was incorporated there was doubt as to the extent to which real mixing with the soil had been achieved, and immobilization might have occurred only in the limited volume of soil immediately adjacent to the straw. The mode of decomposition was also relevant. If breakdown was largely through the action of termites, and this was a strong possibility, then the nitrogen requirement would in any case have been very much less than for fungal or bacterial decomposition.

The second study was that of Odu and Vine (1968b). Working with soils of high C:N ratio from the Nigerian savanna-forest mosaic, they observed that ammonium-N disappeared more rapidly through nitrification than through immobilization. They suggested that, since nitrification involves autotrophic bacteria but immobilization heterotrophic bacteria, the result reflected a lack of decomposable material despite the high C:N ratio. The deficiency of some other nutrient could not, however, be ruled out. Bartholomew (1965), in his review, expressed doubts about the reliability of C:N ratio as an index of immobilization, for it takes no account of the nature of the organic material and the relative exposure of its carbon and nitrogen to microbial action.

Thirdly, there is the work of Blondel (1971c). In a sandy arable soil at Bambey mineral nitrogen from fertilizer applied as a side-dressing to millet disappeared initially more rapidly than could be accounted for by leaching and plant uptake, only to reappear, at least in part, at the end of the season. Soil mineral-N levels (kg/ha) after an application of 200 kg N/ha as ammonium nitrate were:

10 *Sept.*	11 *Sept.*	18 *Sept.*	4 *Oct.*	15 *Nov.*
199	190	67	47	107

This implies substantial immobilization in the absence of any fresh residues. At the same time, when no fertilizer was applied, uptake of nitrogen by millet continued from a soil apparently devoid of mineral-N. The explanation may lie in the vigorous rhizosphere activity of the millet roots, carbonaceous exudates promoting microbial nitrogen fixation when available nitrogen is deficient but immobilization when it is abundant.

These studies show the uncertainty of interpreting results from a tropical context in terms of concepts developed under temperate conditions.

Undoubtedly immobilization occurs, but there is still much to be learned about the factors that govern it and its importance under savanna conditions. In soils in which leaching of nitrate is rapid, a mechanism which conserves fertilizer nitrogen by holding it in organic form and releasing it slowly may be of considerable benefit.

An understanding of immobilization will be particularly important for systems of continuous cultivation, in which the frequent return of crop residues is an essential part of the maintenance of soil structure and fertility. The system advocated in Senegal (Fauck *et al.*, 1969; Charreau and Fauck, 1970) seeks to minimise the effects of immobilization on the subsequent crop by incorporating plant material at the end of the growing season. However, with long-season sorghum as the staple crop, residue incorporation can be carried out only at the beginning of the next season, and the extent to which immobilization then affects an early-planted crop will be an important factor determining the viability of the system. No data are at present available.

Nitrogen may also be immobilized inorganically, through the fixation of ammonium ions in non-exchangeable positions by clay minerals, but the only report of analysis for fixed ammonium is of an alluvial soil from near Lake Chad, Nigeria (Moore and Ayeke, 1965). In this soil the ratio of HF-extractable nitrogen to total (Kjeldahl) nitrogen in the 160 cm deep profile was 0·21 and the profile contained a calculated amount of 1320 kg "fixed" N/ha. There are no data on the mineralogy of this soil, but as fixation requires the presence of 2:1 lattice silicates (Nommick, 1965) this soil was probably atypical of much of the savanna. Fixation is almost certainly much less in well-drained soils where kaolinite is the dominant clay mineral.

(e) *Denitrification*

Denitrification is the biological reduction of nitrate and nitrite to volatile gases, usually nitrous oxide or molecular nitrogen (Broadbent and Clark, 1965). The reduction is carried out by facultatively anaerobic bacteria which can use nitrate in place of molecular oxygen as an electron acceptor. It is relatively independent of nitrate concentration, but is encouraged by locally poor aeration, by high soil pH, by high soil temperatures and by the presence of readily oxidizable organic matter, which both acts as an energy substrate for the denitrifying organisms and also increases the biological demand for oxygen.

Reports of denitrification from savanna areas have been relatively few, and the preponderance of well-drained sandy soils of low organic matter content and acid to neutral reaction suggests that on a regional basis losses are not likely to be serious. However, iron is mobile in many savanna soils, probably as ferrous iron, and where iron is reduced some denitrification is also likely to occur (Patrick and Mahapatra, 1968).

Both Diamond (1937) and Greenland (1956) observed losses of nitrate which could not be accounted for by leaching or plant absorption, and

Greenland (1962) measured large and rapid losses following the addition of nitrate to soil samples from old grassland sites incubated in a saturated state. Losses from unsaturated soils were small, and losses from saturated samples from young grassland and cultivated sites required the addition of a carbonaceous substrate. Nevertheless, the amount of substrate needed was less than that needed to produce an equivalent loss under temperate conditions, and Greenland concluded that this lower carbon requirement and the high soil temperatures could make denitrification losses significant in neutral soils. Losses from long-cultivated soils were likely to be limited by lack of substrate.

McGarity (1962) reported that freezing and thawing of soils stimulated denitrification, a consequence possibly of an increased availability of organic matter. As there is a similar increase in the availability of organic matter when a dry soil is rewetted, the possibility of denitrification losses during the flush at the start of the rains cannot be ignored. This aspect was discussed briefly by Odu and Vine (1968a) with particular reference to the results of Hartley (1961) in Trinidad, who found that flushes of CO_2-production during wetting and drying cycles of a clay soil were accompanied by decreases in mineral-N. And in Uganda, Simpson (1960) found evidence of rapid denitrification of accumulated nitrate when the sudden entry of rain into a dry soil produced reducing conditions.

(f) *Nitrate movement*

Although some soils have positive charges, nitrate is not usually strongly held and very much the greater part of it remains in the soil solution. Thus any downward movement of soil solution involves downward movement of nitrate.

The rate of movement varies widely. First, it depends upon the volume of water moving through the soil. This volume is the difference between rainfall (or irrigation water) and losses from runoff, evaporation, transpiration and changes in water storage. Secondly, over a range of soils the depth of leaching per cm of this drainage water is inversely related to the soil water content at field capacity (Table 26). The relationship is only approximate because there are other factors, but it bears out field experience that light-textured soils lose nitrate more quickly than do heavy-textured soils.

TABLE 26

Relationship between theoretical depth of leaching per cm drainage water and soil moisture content at field capacity (vol %).

Field capacity, %	*Depth of leaching, cm/cm drainage water*
10	10
20	5
30	3·3
40	2·5

TABLE 27

Nitrate leaching rates in various savanna soils

	Site	Soil type	Clay contents, % at different depths, cm				Soil cover	Nitrate source	Movement, cm/cm rainfall
			0—20	20—40	40—60	60—80			
1.	Bambey	Sl. leached Ferruginous	4	7	8	8	Various		7
2.	Sefa	Leached Ferruginous	20	29	32	42	Bare Bare	Native Fertilizer	0·3—0·4 0·4—0·8
3.	Nioro	Leached Ferruginous	9		16		Crop	Fertilizer	1·6
4.	Samaru	Leached Ferruginous	16	27	37	40	Bare	Native	0·5
5.	Samaru	Leached Ferruginous	17	31	40	42	Bare Crop	Fertilizer Native	0·3—0·5 0·2—0·3
6.	Mokwa	Ferrallitic	12	16	27	—	Crop	Fertilizer	1·7

Sources: 1. Blondel, 1971c; 2. Calculated from Blondel, 1971d; 3. Calculated from Blondel, 1971e; 4. Wild, 1972b; 5, 6. Jones, 1975a and unpublished data.

Field measurements of leaching rates have been relatively few. At Samaru, by field sampling, Wild (1972b) obtained a value of 0·5 cm per cm rain for the rate of leaching of nitrate mineralized and nitrified at the start of the rainy season in bare-fallow plots. Blondel (1971c, d, e, f) made similar measurements at three sites in Senegal. At all three, under bare fallow, native or fertilizer nitrate was rapidly leached below 20 cm; but for deeper leaching there were big differences between sites. In the sandy Bambey soil (3 per cent clay) nitrate was carried below 100 cm in both years of measurement, when annual rainfall was 532 and 823 mm. On a leached Ferruginous Soil at Sefa substantial amounts of nitrate remained at a depth of 80 cm at the end of one year when rainfall was 1254 mm, but in another season little remained in the top 100 cm after 1439 mm rain. In another leached Ferruginous Soil at Nioro-du-Rip the nitrate concentration remained high at a depth of 80—100 cm after 952 mm rainfall.

Blondel (1971c) calculated the rate of leaching at Bambey to be 7 cm per cm rainfall, but on all the leached Ferruginous Soils investigated rates were very much lower (Table 27). This difference is too great to be explained solely by differences in water content at field capacity due to higher clay content. In relation to the calculated drainage, the leaching rate at Samaru was much less than that predicted from field capacity (Wild, 1972b). The probable explanation lies in the physical properties of the soil and in the intensity of the rains. Leached Ferruginous Soils with well developed textural B horizons often contain cracks and root channels, and a high proportion of the rain falls in storms of high intensity. It seems likely that leaching is slow in these soils because much of the drainage water from this high intensity rainfall passes through the cracks and channels of the B horizon too quickly to allow nitrate held in finer pores to diffuse into it.

The Bambey soil provides a contrast. Blondel (1971c) reported that the maximum nitrate concentration in the drainage water of the Bambey lysimeters (40 cm deep) occurred after a cumulative drainage of only 33-34 mm, less than that expected from the water content at field capacity (40 mm). This phenomenon of early displacement is to be expected from a sandy soil where there are few fine pores and no large aggregates, but where there are "dead pores" which do not conduct drainage water.

Where leaching is as rapid as it is at Bambey, some advantage may be gained by applying fertilizer-N in ammonium form, as lysimeter results show (Table 28).

In the absence of fertilizer nitrogen, actual losses of nitrate from savanna soils vary with the vegetative cover. Under natural vegetation a number of factors combine to keep them to a minimum:—

(a) Vegetative cover is generally good, so that high rates of transpiration reduce percolation and, hence, leaching. Moreover, where the flora includes perennial species, transpiration may go on through a large part of the dry season, building up a substantial moisture

TABLE 28

Nitrogen losses from soil under millet in Bambey lysimeters (Blondel, 1971a).

	Control	*300 kg N/ha applied at planting:* as KNO_3	as $(NH_4)_2SO_4$
Losses of N as kg/ha:	3·9	69·4	9·9
as % fert-N:	—	23·1	3·1

deficit in the soil and so delaying the start of deep leaching at the beginning of the next wet season.

(b) Biological processes and especially uptake by the vegetation tend to keep levels of soil nitrate-N low.

(c) The existence of perennial vegetation means that there is already an extensive network of roots present in the soil at the beginning of the season able to intercept efficiently the "flush" of nitrate moving down.

These mechanisms of conservation are much less pronounced under crops. The results of Tourte *et al.* (1964) illustrate the effect that the nature of the vegetative cover has on leaching losses. Measured losses of nitrogen from lysimeters 40 cm deep were, according to cover:—

A millet green manure crop	1·6 kg/ha
Natural fallow	6·4 kg/ha
Bare fallow	45·6 kg/ha
Groundnuts after green manure	14·9 kg/ha
Groundnuts after natural fallow	3·8 kg/ha

In this experiment the natural fallow was recently established and would have had few if any perennials of any size. Comparison with bare fallow, however, shows how important it is to have some form of vegetative cover, if only weeds, to conserve the nitrogen.

Three other aspects of nitrate movement must be mentioned:—

(i) The leaching of nitrate cannot be divorced from the leaching of cations. As Nye and Greenland (1960) stressed, the total concentration of cations in the soil solution depends on the total concentration of anions, and of these the nitrate ion, and to a lesser extent, the bicarbonate ion are quantitatively the most important. Any increase in the leaching of nitrate involves a corresponding loss of cations and an increase in soil acidity.

(ii) Many savanna soils have an accumulation of clay in the subsoil. Wild (1972b) found the positive charge in the textural B horizon to be only 0·2 me/100 g, but even this low value could give an adsorption of up to 28 ppm NO_3-N (the actual amount depending on the concentrations of other anions). In soils of low pH and containing hydrated iron and aluminium oxides of high surface area, positive charges will be much greater and nitrate adsorption can be higher. Kinjo and Pratt (1971) found a significant correlation between nitrate adsorption and the content of amorphous materials in subsoils from Mexico and South America. Adsorption from 0·2 M nitrate solution was greater in a Ferrallitic Soil (oxisol; 2—6 me per 100 g soil) over the pH range pH 5·0—3·0 than in a Ferruginous Soil (alfisol: 1 me per 100 g soil).

Leaching of nitrate out of the topsoil does not therefore necessarily mean permanent loss. Simpson (1961), working on a free-draining red earth at Kawanda, Uganda, noted that while there was considerable leaching of nitrate during the arable phase of a rotation much was later recovered by perennial grasses and some cereal crops; and Jones (1968) at Namulonge, Uganda, showed that the greater part of the nitrogen gained by the soil under fallow vegetation came from the uptake of nitrate leached into the subsoil during cropping.

(iii) In northern Australia, dry season accumulations of nitrate in the topsoil were ascribed to upward movement of subsoil nitrate as the topsoil dried out (Wetselaar, 1961). However, the greater part of a similar accumulation in Uganda was microbiological in origin, and capillary transport or diffusion played only a minor role (Simpson, 1960; Stephens, 1962). At Samaru, Wild (1972b) found no dry season accumulation of nitrate in the topsoil, and it is thought that upward movement of nitrate is unlikely to be of any great significance in savanna agriculture.

(g) *Management of soil nitrogen*

Lack of available nitrogen is the nutritional factor which, with phosphate deficiency, most severely limits crop yields. Although aggravated in the first year or two after clearing from fallow by immobilization, this lack arises primarily from low total nitrogen in the soils. This is particularly so in soils that are frequently or continuously cropped and in drier areas where the climate and the sandy texture of the soils militate against any build-up of soil organic matter. Careful management and the maximum return of organic residues may effect a considerable improvement, but it seems unlikely that the nitrogen demands of high-yielding cereal crops will ever be met solely by the native soil nitrogen mineralized during the season of growth.

A partial answer may lie in making greater use of the legume phase of the crop rotation. Comparatively little attention has yet been given to the contribution of groundnuts and cowpeas to the nitrogen supply of subsequent crops or, where intercropping is practised, of companion crops. Even where all above-ground residues are removed from the field a considerable benefit is possible. Thus in a recent experiment at Samaru, where no nitrogen fertilizer was applied maize following groundnuts took up 30 kg/ha more nitrogen, and yielded 1000 kg/ha more grain, than maize following sorghum; but no similar benefit accrued from cowpeas, and it is not known why (Jones, 1974). With the rapidly rising cost of nitrogen fertilizer, it is to be hoped that agronomists, plant-breeders and microbiologists will recognize the urgent need that exists for increasing the contribution of symbiotic fixation to the nitrogen economy of the whole cropping cycle.

One of the objections to the inclusion of fallows in crop rotations is that they are non-productive. The same objection does not apply to leys and pastures within a mixed farming system, and these can effect a greater improvement in soil organic matter and nitrogen status. In Australia legume-based pastures are effective in maintaining and improving the fertility of wheat soils, and subsequent wheat yields are related to the accumulation of organic nitrogen (Greenland, 1971). The feasibility of using ley-arable rotations to maintain the fertility of savanna soils is also in urgent need of investigation.

Even so, it seems inevitable that much of the increased demand for nitrogen that is posed by more intensive crop production will be met by the increased use of inorganic fertilizers, and this is a development that has its dangers. The most widely used source of fertilizer nitrogen in West Africa has until now been ammonium sulphate, and its serious acidifying effect on the poorly buffered savanna soil is well documented (Stephens, 1960b; Bache, 1965; Poulain, 1967; Bache and Heathcote, 1969; Charreau and Fauck, 1970; Roche, 1970). Calcium ammonium nitrate and urea, the two most suitable alternatives, have certain drawbacks; but these must be overcome, for high productivity will be maintained in the long term only if acidification is contained. We return to this subject in Chapter 10.

Chapter 7

PHOSPHORUS

Deficiency of phosphate occurs widely in the savanna, and in some soils deficiency is so acute that plant growth virtually stops as soon as the seed reserves of phosphate have been used. In this chapter we examine the phosphate status of the soils of the region, the chemical forms in which soil phosphate occurs and the factors affecting its availability to plants. Crop responses to phosphatic fertilizers will be discussed later (Chapter 10).

1 TOTAL PHOSPHORUS

The total amounts of phosphate in the soils are generally low and sometimes very low. In 334 topsoil samples (Table 7) total contents ranged from 13 to 630 ppm P, with a mean of 143 ppm. Amounts were highest in Vertisols and Hydromorphic Soils (means 194 and 182 ppm, respectively) and lowest in Brown and Reddish-Brown Sub-Arid Soils (mean 92 ppm). Further data collected in Table 29 also show low total contents. The only exceptions are the two samples of soils derived from basalt in Cameroun, in which amounts were 375 and 1087 ppm P.

TABLE 29

Total phosphorus contents of some savanna topsoils

Source	*Country*	*Soils*	*No. of samples*	*Total P (ppm) Range*	*Mean*
1	Senegal, Niger, Dahomey, Upper Volta.	Ferruginous	20	24—226	83
	Senegal, Guinea, Togo, Dahomey	Ferrallitic	7	19—370	149
	Cameroun	Ferrallitic/basalt	1	—	1248
	Togo, Cameroun	Black clay	2	149—277	213
	Mali	Sub-arid Brown	2	92—116	104
	Cameroun	Brown soil/basalt	2	375—1087	731
2	N. Nigeria	Unfertilized farmers' plots	54	—	107
3	Ghana	High-grass savanna	67	—	134
4	Nigeria	Savanna	14	—	110

Sources: 1. Bouyer and Damour (1964); 2. Goldsworthy and Heathcote (1963); 3. Nye and Bertheux (1957); 4. Enwezor and Moore (1966a).

Combining all these analyses gives a mean total content of 140 ppm P in 503 samples of topsoil. This figure is probably a little low for analytical reasons, because the phosphorus in many samples was determined by digesting the soil with hot concentrated acids such as perchloric acid, but the correction factor is likely to be small. The value of 140 ppm P is substantially lower than means reported from Australia (350 ppm in 2217 samples) and the United States (568 ppm in 122 samples) by Wild (1958), and lower than almost all the values collected from seven countries by Nye and Bertheux (1957); only some Malayan soils had a lower content.

No pedological study has been done to account for the low phosphate contents, and therefore any explanation is at present speculative. The factors discussed below seem the most relevant.

(a) *Effect of parent material*

Average rock analyses given by Rankama and Sahama (1950) and Goldschmidt (1962) were 870 to 2440 ppm P for igneous rocks, with granites the lowest, and 175 to 1310 ppm P for sedimentary rocks, with sandstones the lowest; basic igneous rocks were highest in phosphate. Analyses quoted by Nye and Bertheux (1957) of rock samples from Ghana showed an average of 700 ppm P for fifteen samples of granite, and 750 to 2440 ppm P for rocks of intermediate and basic composition. Few outcrops of basic rocks occur in the savanna, and the mean phosphate composition of parent rocks is probably below the world average, though there are no representative analyses.

Differences between parent rocks are reflected in soil phosphate contents (Tables 30 and 31). Acquaye and Oteng (1972) divided soils from Ghana into two groups with respect to total phosphate content: (a) soils developed over granite and acidic gneisses, sandstones and Tertiary sands (i.e. acidic parent material) which had a mean topsoil content of 139 ppm P, and (b) those over basic rocks (limestone, hornblende gneiss), Birrimian rocks, phyllite and alluvium, which had a mean content of 232 ppm. High amounts found in soils formed on clay deposits (Table 30) and alluvium may have been due to the clay holding phosphate against leaching. Total soil phosphate was highly correlated with clay content in soils on alluvium and in soils on sandstones and sands, but not at all in soils over igneous rocks (Acquaye and Oteng, 1972). However, while parent material clearly influences soil composition, it is improbable that it is the only cause of the predominantly low phosphate contents of savanna soils.

(b) *Loss by leaching and erosion*

In Ghana, Nye and Bertheux (1957) suggested that leaching was an important mechanism of phosphate loss. Except where the soil is sandy, annual movement out of the upper part of the profile is probably small because surface reactions with clay silicate minerals and free iron and aluminium oxides keep soil solution concentrations low; but much phosphate may be lost from the zone of rock weathering (Wild, 1961), where phosphate solubility appears to be higher. From the top 40 cm of a soil from Bambey, Vidal and Fauche (1962) reported an annual loss of 0·1 kg P/ha.

Little is known of the age of the soil material; land surfaces are believed to date from the mid-Tertiary, and although the soil is not necessarily of the same age, there is evidence that it has been exposed to climatic periods more humid than at present. The extensive occurrence of plinthite implies that the iron was mobile probably as ferrous iron, resulting in an increased solubility of phosphate, as occurs during the flooding of paddy soils. The present low content of soil phosphate may therefore be a consequence of the age of the soil material and the pedogenetic processes to which it has been subject in the past.

Loss of topsoil through erosion involves loss of phosphate; but in some areas erosion keeps profiles shallow, so that weathering rock is within range of the roots of the natural vegetation, and topsoils are correspondingly enriched. Elsewhere, wind-blown sand may give a surface deposit low in phosphate.

(c) *Effect of vegetation and climate*

In general, soil phosphate contents are higher in the forest zone than in the tall-grass savanna (Table 31), and are lowest in the Sub-Arid Soils of

TABLE 30

Total phosphorus in savanna topsoils according to parent material (References as in Table 7).

Parent material	No. of samples*	Total P (ppm) Mean	Range
Aeolian deposits	59	82	13—290
Clays: lacustrian and alluvial	14	254	66—591
Other alluvium	44	149	31—510
Sandstones	35	143	42—560
Sands and other unconsolidated sediments	85	143	38—580
Granite	43	134	29—420
Schist	44	133	33—230
Gneiss-hornblende	3	246	207—306
Gneiss-other	13	103	30—210
Undifferentiated Basement Complex	19	141	92—434
Basalt	4	325	160—630
Ironpan and other ferruginous deposits	25	142	75—264

* Generally from 0—15 cm.

TABLE 31

Total and organically combined phosphorus in surface (0—15 cm) soils of Ghana (Nye and Bertheux, 1957).

Parent material	No. of sites	Total P, ppm	Org.P, ppm	Org.P/total P	C/org. P	N/org. P
Forest region:						
Granite, gneiss, schist	11	266	86	0·32	251	24·5
Sandstone, sandy clays	4	215	61	0·28	295	22·5
Basic igneous or metamorphic	6	509	149	0·29	160	15·9
Mean	21	326	99	0·30	233	21·6
Savanna region:						
Granite, gneiss	15	70	18	0·25	267	27·5
Sandstone	46	130	26	0·20	230	16·8
Basic igneous or metamorphic	6	325	60	0·18	292	21·8
Mean	67	134	27	0·21	247	19·5

the north (Table 7). These differences follow the same pattern as those of organic carbon and total nitrogen (Acquaye and Oteng, 1972) and appear to reflect the extent to which the natural vegetation returns phosphate to the soil surface in litter and dead roots. Nye and Greenland (1960) found that the vegetation of a 40-year-old forest contained about 120 kg P/ha whereas that of a 20-year-old savanna only about 14 kg P/ha. The return to the soil surface can be expected to decrease as the vegetation becomes sparser in the drier parts of the savanna, and this may account for why the few analyses of savanna soils so far published show no consistent accumulation of phosphate at the surface (Nye and Bertheux, 1957; Enwezor and Moore, 1966a; Ipinmidun, 1972).

There may be a loss of phosphate when the vegetation is burned. It is known that a loss occurs when some types of vegetation are dry-ashed in the laboratory unless a suitable calcium or magnesium salt is added. There appear to be no data from a burn in the field.

2 FORMS OF PHOSPHATE

(a) *Organic*

Estimates of the proportion of total phosphate in organic form in topsoils have varied widely. Nye and Bertheux (1957) reported 21 per cent for soils of the high grass savanna as compared with 30 per cent in the forest samples, but more recent analyses from Ghana gave 45 per cent for savanna and 51 per cent for forest soils (Acquaye and Oteng, 1972). Some values from savanna soils in northern Nigeria were intermediate: means of 36 per cent for 54 soils (Goldsworthy and Heathcote, 1963) and 39 per cent for 12 soils (Ipinmidun, 1972). However, in long-cultivated unfertilized soils at Samaru, Kano and Yandev organic phosphorus percentages were only 17, 22 and 22, respectively (Bache and Rogers, 1970), which is consistent with a decline in soil organic matter during long periods of cultivation. Enwezor and Moore's (1966b) mean value of 25 per cent for four savanna soils was also low. One may conclude only that values are rather variable and probably depend on vegetation and site history and, not least, on the analytical methods used (Enwezor and Moore, 1966a). However, it is clear from the profile data in several of the above-mentioned publications that amounts of organic phosphorus decline with depth, though not necessarily as rapidly as carbon.

There is no information on the chemistry of organically-combined phosphate in savanna soils, but under forest in Nigeria between 23 and 30 per cent was present as inositol phosphates, a proportion similar to that in three uncultivated English soils (Omotoso and Wild, 1970).

(b) *Inorganic*

Fractionation of inorganic phosphate by selective chemical extraction, following the methods of Chang and Jackson (1957), has been reported by Dabin (1963, 1967), Bouyer and Damour (1964) and Enwezor and Moore (1966b) among others (see review by Pichot and Roche, 1972). A large proportion is found occluded in ferruginous coatings and concretions (Nye and Bertheux, 1957; Bouyer and Damour, 1964; F.A.O., 1967). Table 32 gives one example.

According to the Chang and Jackson method, the form of the non-occluded inorganic phosphate shows a dependence on soil pH. It is associated

TABLE 32

A chemical fractionation of soil phosphorus (Bouyer and Damour, 1964).

Soil type	*No. of samples*	*Mean P, in p.p.m.* non-occluded			occluded	
		Al—P	*Fe—P*	*Ca—P*	*Al—P*	*Fe—P*
Ferruginous	10	15	24	11	6	124
Slightly Ferrallitic	5	47	27	9	10	74

with calcium and aluminium in neutral soils, increasingly with aluminium down to pH 5·5, and with iron, and probably organic matter, under more acid conditions (Pichot and Roche, 1972).

Phosphate-rich concretions may be present in the sand and gravel fractions, as found by Nye and Bertheux (1957) in a forest soil, but in the absence of these concretions the phosphate appears to be concentrated in the clay fraction (Ipinmidun, 1972).

3 REACTIONS OF ADDED PHOSPHATE IN SOIL

From a review of field experiments carried out in the tropics Russell (1968a) showed that it is quite wrong to describe tropical soils as generally having high phosphate fixation capacities in the sense that they rapidly convert applied phosphate into unavailable forms. In the savanna, crop responses have often been obtained from very small applications of phosphate, of the order 4—10 kg P/ha (Nye, 1954; Braud, 1967; Goldsworthy, 1967b; F.A.O., 1968); and experiments in Ghana (Nye, 1954; Stephens, 1960b), northern Nigeria (Greenwood, 1951; Goldsworthy, 1968), Dahomey (Thevin, 1967), Mali and Ivory Coast (Pichot and Roche, 1972) have all shown a pronounced residual effect from a single application of phosphate fertilizer. One example is given in Table 33.

The reactions of fertilizer phosphate with soil are now becoming better understood. Close to the fertilizer particles or granules, and diffusing from them, the solution has a very high phosphate concentration and low pH. In this zone precipitation of dicalcium phosphate and other products of intermediate and low solubility occurs (Huffman, 1962). There are further reactions because these products have a higher solubility than the equilibrium soil/phosphate system. Although equilibrium is only very slowly if ever achieved, the further reactions include precipitation of basic calcium phosphates, e.g. calcium octophosphate and apatite, in soils above pH 7 and, more generally, adsorption on to the surfaces of soil minerals including clay minerals, hydrous metal oxides and carbonates (Larsen, 1967). In acid

TABLE 33

Effect of phosphatic fertilizer in the year of application and in two subsequent years, N.E. Dahomey (Thevin, 1967).

Fertilizer in first year, kg/ha		*Crop yields (kg/ha)*	
	1st year cotton	*2nd year groundnuts*	*3rd year* sorghum*
nil	765	1200	786
75 ammonium sulphate	934	1255	824
75 ammonium sulphate + 100 triple superphosphate	1200	1710	1042
75 ammonium sulphate + 110 dicalcium phosphate	1261	1660	1234

* 40 kg/ha ammonium sulphate given to plots previously fertilized.

and neutral soils adsorption occurs by an exchange reaction with hydroxyl ions coordinated with iron and aluminium, or by exchange with hydroxyl ions dissociated from water molecules held by the same metal ions (Hingston *et al.*, 1968). Except where soil pH is above 7·0, most phosphate added to savanna soils appears to be linked to aluminium, at least initially, although some iron linkages are formed in very acid soils (Dabin, 1970; Pichot, Truong and Burdin, 1973).

Some soils, for example those formed from recent volcanic ash or lavas high in iron and aluminium, have mineral constituents with highly reactive surfaces. These constituents include allophane, amorphous hydrated oxides of iron and aluminium and, to a lesser extent, crystalline iron and aluminium oxides and layer lattice silicate minerals (Younge and Plucknett, 1966; Fox *et al.*, 1968; Le Mare, 1968). Adsorption capacities of the more common constituents of the clay fraction have been rated : amorphous hydrated oxides>gibbsite = goethite>kaolin>montmorillonite (Fox *et al.*, 1968).

Most savanna soils are unlikely to contain high proportions of reactive minerals. Ferrallitic Soils in particular may contain quite large amounts of iron and aluminium oxides, but mostly in a relatively inactive, crystalline form, of lower surface area than the non-crystalline gels of volcanic soils (Greenland, 1972). Exceptions are provided by some Hydromorphic Soils (Jacquinot, 1965; Anon, 1967, p. 434) especially when acid; soils formed from recent volcanic ash and lavas, as in Cameroun; and possibly some highly leached acid Ferrallitic Soils.

There is also the evidence of laboratory experiments that phosphate adsorption by savanna soils is moderate to low. Bouyer and Damour (1964), working on a range of soils from Dahomey, Guinea, Niger, Senegal, Togo and Upper Volta, showed that adsorption varied with soil type:—
Hydromorphic (2 samples)> Ferrallitic (4)> Ferruginous (13), although they gave no details of method of measurement. Bhat and Bouyer (1968)

showed that a heavy Hydromorphic Soil from the Senegal River area had a very much higher adsorption capacity than a sandy Bambey soil, 1060 ppm as against 62 ppm.

In Nigeria, Enwezor and Moore (1966b) shook samples from nine forest and savanna profiles with 150 μg P per g soil and found that in most surface samples less than half the added phosphate was adsorbed in 24 hours. In their four savanna profiles adsorption appeared to be related to clay content and in three of them it increased with depth. The relationship between adsorption and clay content was also demonstrated by Udo and Uzu (1972), who showed that in a range of soils from eastern and western Nigeria (forest and southern savanna) adsorption was positively and significantly correlated with clay content and exchangeable Fe and Al and negatively and significantly with pH. In general, adsorption capacity was greater in the more strongly weathered (mostly Ferrallitic) soils of eastern Nigeria than in the less strongly weathered (Ferruginous) soils of the west.

However, adsorption should not be confused with fixation in unavailable form. As long as adsorbed phosphate remains exchangeable, that is, it can undergo isotopic exchange, it is available to plants (Larsen, 1967; Juo and Ellis, 1968). Adsorbed phosphate may eventually be converted into non-exchangeable form (Larsen *et al.*, 1965), but there is at present no information on the rate of this reaction in savanna soils.

4 AVAILABILITY OF SOIL PHOSPHATE TO PLANTS

(a) *Estimation by extraction procedures*

In a region in which phosphate deficiency is a major factor limiting crop growth, much importance attaches to the assessment of soil phosphate availability; but attempts to find a suitable availability index by *ad hoc* extraction procedures have usually been unsuccessful (Nye and Bertheux, 1957; Homer, 1962; Dabin, 1963). For instance, comparisons between six extraction procedures in Ghana showed that, while good correlations with crop yield occurred sporadically with most methods, none was suitable anywhere in all years (Halm, 1968). In an area of fairly uniform soils in Senegal, Ollagnier and Gillier (1965) obtained a significant correlation ($r = 0{\cdot}366^{**}$) between crop response to phosphate fertilizer and soil phosphate extractable with hot nitric acid, but there was a great deal of variability at individual sites which remained unaccounted for.

Undoubtedly, some of the variation in crop response to phosphate fertilizers arises from deficiencies in other nutrients, from differences in water supply between years and between sites, and from varying standards of husbandry. In pot experiments these effects can be overcome and differences between soils more easily recognised. Oteng and Acquaye (1971) compared ten extraction methods on 48 representative soils in Ghana. They examined relationships between extractable phosphate and the uptake and response to phosphate of millet in pots, and the influence of soil and site factors on these relationships. Four of their methods justified testing against crop performance in the field, although they appeared less promising with soils from the savanna than from the forest.

The empirical extraction approach is open to criticism because phosphate uptake by plants is controlled by intensity, quantity and rate factors. Not only do the magnitudes of these vary widely from place to place according to soil type and site history but also their relative importance differs between crops. It remains to be seen whether any empirical method will provide a serviceable index, that is, one that will pick out reliably those soils where phosphate is needed and, at least approximately, indicate the amount of fertilizer required. Oteng and Acquaye (1971) discussed the interaction between the chemical nature of the extractant and the chemical form of the soil phosphate; and Dabin (1971), related available phosphate, as measured by a modified Olsen technique (Dabin, 1967), to changes in the forms of inorganic phosphate in the soil under different manurial regimes. It seems likely that *ad hoc* extraction techniques will be chosen successfully only where there already exists a knowledge of the predominant forms of phosphate present in the soil, their relative availability to crops and their solubility in the available extractants. The success of empirical extraction methods depends finally on how well they correlate with field response to fertilizers. Results to date show that a great deal more work is needed, and that techniques found useful in temperate areas should not be applied uncritically in the savanna.

Only one instance will be quoted of the theoretically sounder quantity/intensity approach to available phosphate. Bache and Rogers (1970) related various measures of phosphate availability to the phosphate uptake and dry matter yield of Rhodes grass (*Chloris gayana*) grown on soils from long-term manurial trials in Nigeria and showed the importance of both quantity and intensity factors to the phosphate supply. The inorganic phosphate extracted by HCl and NaOH (a quantity measure) combined with the logarithm of phosphate concentration in a saturated water extract (an intensity measure) accounted for 91 per cent of the yield variation. However, the best single measure was given by resin-extractable phosphate (a composite quantity/intensity parameter), and, as these authors suggested, its relative simplicity recommends its adoption for advisory purposes.

(b) *Soil factors*

Phosphate availability is determined by the concentration maintained at the root surface. It therefore depends on (i) the solution concentration before phosphate is taken up by roots (intensity factor), and (ii) on the maintenance of this concentration, which in turn depends on desorption of labile phosphate (quantity factor) from soil surfaces, and possibly on mineralization of organically-bound phosphate. These processes have been described by Fried and Broeshart (1967), Larsen (1967) and others, and will be discussed here only in relation to work on savanna soils.

(i) Intensity factors: Concentrations of phosphate in soil solutions, at least of unmanured soils, are very low. Bhat and Bouyer (1968) reported concentrations in 0·005 *M* $CaCl_2$ extracts (soil: solution ratio of 1:4) of 0·01 and 0·06 mg P/l in a Hydromorphic and a sandy Ferruginous Soil, respectively, from Senegal. Similarly low concentrations (0·01—0·02 mg P/l) were found in saturation extracts of three northern Nigerian Ferruginous Soils (Bache and Rogers, 1970).

(ii) Quantity factors: Labile phosphate, that is, the pool of readily desorbed phosphate can be estimated by isotopic dilution methods. Bang and Oliver (1971) gave the following ranges for 24-hour isotopic exchange in soils of three different types in Madagascar:

Ferruginous:	4·4—26 ppm
Ferrallitic:	15—120 ppm
Hydromorphic:	50—200 ppm

These figures conform with the few values we have for savanna soils. The Hydromorphic Soil, described by Bhat and Bouyer (1968), contained 75 ppm of isotopically-exchangeable P and required large amounts of phosphate to give a crop response because of its high adsorption capacity. By contrast, the Ferruginous Soil contained only 8·7 ppm of isotopically-exchangeable P and, having a low phosphate sorption capacity, required only small amounts of fertilizer. Phosphate added to the Ferruginous Soil in short-term laboratory experiments remained isotopically-exchangeable, whereas in the Hydromorphic Soil between 34 and 67 per cent became non-labile. Three unmanured Ferruginous Soils of Bache and Rogers (1970) contained between 7 and 11 ppm of isotopically-exchangeable P (Table 34). These amounts are low and may be contrasted with 50—150 ppm in 24 unfertilized soils in U.K. (Larsen *et al.*, 1965).

In the three Nigerian soils, previous phosphate fertilization raised both the concentration of soluble phosphate and the amount of labile phosphate; increases in soluble phosphate were particularly large in the Kano and Yandev soils (Table 34). In general, the effect of fertilizers varies with adsorption capacity. Where this is high, fertilizer phosphate tends to increase the labile pool but has only a small effect on soluble phosphate concentration; but where adsorption capacity is low, additions of phosphate raise soluble concentration very considerably.

As shown above, savanna soils on well drained sites have low to moderate phosphate adsorption capacities, and hence phosphate manuring increases solution concentration. Whether this increase is sufficient depends on the amount of fertilizer used and on the rate of demand by the crop. Short-season, high-yielding crops have a higher rate of demand than traditional

TABLE 34

Parameters of available phosphate on plots given no fertilizer, superphosphate or dung, in three long-term experiments, northern Nigeria (Bache and Rogers, 1970)

Soil treatment and site	*Quantity measurements, ppm soil*		*Saturation extract, (P) × 10⁻⁶ moles/l*	*Pot experiment: P uptake,* mg P/kg soil*
	E value	*L value*		
Control — Samaru	8·1	12·0	0·23	3·9
Kano	10·9	9·2	0·57	1·4 (4·6)
Yandev	7·3	11·8	0·31	0·7 (4·8)
means	8·8	11·0	0·37	(4·4)
[1]Superphosphate—				
Samaru	11·8	25·4	0·77	11·9
Kano	16·7	25·2	10·75	8·0 (17·4)
Yandev	14·2	15·9	3·20	3·1 (13·0)
means	14·2	22·2	—	(14·1)
[2]Dung— Samaru	7·6	22·8	0·48	4·6
Kano	9·2	16·5	1·86	3·0 (8·9)
Yandev	8·5	12·5	0·86	0·9 (5·4)
means	8·4	17·3	1·07	(6·3)

[1] 9·8 kg P/ha/year as single superphosphate for 15 years.
[2] 5 tonnes dung/ha for 15 years, per cent P not known.
* Figures in parentheses are estimates of P uptake under optimal growing conditions.

crop varieties and need a higher concentration and more fertilizer applied per hectare. New field experiments are therefore required whenever an innovation is introduced (e.g. better crop husbandry, improved crop variety), in order to assess the optimum fertilizer rate and frequency of application. Such experiments are expensive and often difficult to carry out well, and it would be useful to supplement them with measurements to characterize the soils. Quantity and intensity factors, that is, isotopically-exchangeable and soil solution phosphate concentration, are promising where facilities exist for their measurement. Alternatively, one might measure the degree of phosphate saturation of the soil (Woodruff and Kamprath, 1965), the resin-extractable phosphate (Bache and Rogers, 1970) or the phosphate addition needed to raise the soil solution to a given concentration, say 0·2 ppm P (Barrow, 1973); but all these methods require further investigation before they can be useful.

(c) *The organic matter contribution*

Part of the phosphate supply to plants comes from the mineralization of organically-bound phosphate, though amounts are probably quite small. Organic phosphate levels in most savanna soils are low, lying predominantly in the range 15—60 ppm (Table 31; Goldsworthy and Heathcote, 1963), and rates of mineralization are uncertain. Although Acquaye (1963) found evidence of substantial mineralization in forest soils in Ghana, and Friend and Birch (1960) obtained a significant correlation between organic phosphorus and crop response to phosphate fertilizer in Kenya, correlations between soil organic phosphate and plant uptake of phosphorus have usually been low and non-significant (Nye and Bertheux, 1957; Homer, 1962; Bache and Rogers, 1970). This implies that organically derived phosphate makes only a small contribution to plant uptake in these soils, although its contribution might well be masked by variations in the inorganic phosphate supply.

Wide organic C: organic P ratios (mean: 247) have been reported from the southern savanna (Nye and Bertheux, 1957), but the lower values (means: 114 and 119) found in northern Nigeria (Goldsworthy and Heathcote, 1963; Ipinmidun, 1972) are more similar to ratios elsewhere (Barrow, 1961). Nye and Bertheux (1957) found no significant change in the C:P ratio on cultivation, suggesting that organic phosphorus was mineralized at about the same rate as the organic matter as a whole. As much of the phosphate is contained in the biomass, the factors involved in its mineralization (Birch, 1961) are likely to be similar to those involved in nitrogen mineralization (soil pH, wetting/drying effect, etc) and hence may be partly responsible for the poor correlations reported above.

If one accepts an annual organic matter decomposition constant of 4 per cent for cultivated savanna soil (Chapter 5), 15—60 ppm of organic phosphorus release only 2·4—10 kg P/ha to a depth of 30 cm, and not all of it will be taken up by the crop. But at low levels of fertility even this amount may be important.

There is some evidence of another interaction between soil organic matter and phosphate availability. Bhat and Bouyer (1968), Bhat (1970) and Oliver (1972) have reported that the application of humified organic matter or manure to several different soil types from Senegal and Madagascar (Ferruginous, Hydromorphic and Ferrallitic) increased (i) phosphate concentration in the soil solution, and (ii) labile phosphate, as measured by isotopic dilution. Pichot, Truong and Burdin (1973), in Central African Republic, also reported that manure increased soluble phosphate. The probable cause is competition between humate and phosphate ions for adsorption sites. In pot experiments, additions of organic matter have been shown to benefit the growth and phosphate uptake of ryegrass (Pichot and Roche, 1972), but the significance of this effect at field-crop level remains to be demonstrated.

(d) *Management of soil phosphate*

A very large part of the savanna is covered by Ferruginous and Sub-Arid Soils, in which both soluble and labile phosphate levels are low; but because adsorption capacities are also low, small additions of fertilizer phosphate increase soluble phosphate and give a significant crop response. Moreover, the conversion of adsorbed phosphate into fixed, unavailable form appears to be generally very slow. Hence, applied phosphate remains available over several seasons and total recoveries are high, although more experiments are needed to determine the optimum frequency of application.

Similar remarks apply to many of the Ferrallitic Soils found in the savanna; but in some intensely weathered Ferrallitic Soils and in some Hydromorphic Soils, the presence of large amounts of surface-active minerals means that adsorption capacities are high. Here, large amounts of phosphate fertilizer are required to raise the levels of soluble phosphate and give crop responses. Although little experimental work has been done with these soils, it might prove best to maintain high levels of organic matter, to lime to pH 7 where necessary (Pichot and Roche, 1972), and to use fertilizer placement or banding for all crops. If cheaply available, silicate might be useful in helping to saturate the phosphate adsorbing sites.

Chapter 8

OTHER NUTRIENTS

1 SULPHUR

(a) *Amounts and forms*

Compared with the soils of many temperate and forested tropical areas (Whitehead, 1964), the total sulphur content of most savanna soils is low. Goldsworthy and Heathcote (1963) reported a range of 18—132 ppm (mean 45 ppm) for 54 topsoils (0—15 cm) from northern Nigeria. Although no analyses have been reported, the acid sulphate soils of the Senegal river delta and gypseous soils in dry northern areas can be assumed to be exceptions to this rule.

As Whitehead pointed out, total sulphur contents are usually fairly closely correlated with organic matter and nitrogen contents, and in the area sampled by Goldsworthy and Heathcote total sulphur and organic sulphur increased from north to south, following the trend of the organic carbon, total nitrogen and total phosphorus values. The C:S ratio (mean: 94) increased in the same direction. More recently, Cooper (1971) fractionated the sulphur in soils from the same area, and his results confirm that the sulphur is largely in organic forms (Table 35).

Organic sulphur decreases with depth, but its distribution in the profile is determined very much by the rooting pattern of the previous vegetation. Bromfield (1972), working at Samaru, showed that the root remains from natural fallow, present up to nine years after clearance, strongly influenced the distribution of organic sulphur; but after 19 years the distribution mainly reflected the crop rooting pattern.

Bromfield also demonstrated the dependence of soil inorganic sulphur on the type of land use. Under natural fallows amounts were small (2—3 ppm) and varied little with depth. Where fallows had been cleared but not cropped 10—20 ppm sulphate-S had accumulated between 10 and 30 cm depth, but on fields cropped for long periods without fertilizers amounts were again small at all depths. The concentration of adsorbed sulphate just below the topsoil in uncropped soils probably represents the accumulation over many years of sulphate mineralized from soil organic sulphur and deposited from the atmosphere. In cropped and bush fallow soils this had been efficiently intercepted by plant roots.

Where cropped soils had regularly received fertilizers containing sulphur there was again an accumulation, generally of much greater magnitude and extending to greater depths. The implication that the soils held sulphate-S very strongly against leaching is supported by Bromfield's sulphur balance sheet for a 19 year-old fertilizer experiment (Table 36). When allowance is made for crop removal (48 and 86 kg S/ha on treatments 2 and 3), losses of applied sulphate by erosion and leaching were probably less than 10 per cent.

Sulphate adsorption capacity depends largely on clay content and pH, and differences may therefore be expected between leached Ferruginous Soils with high concentrations of kaolinite in the subsoil (as at Samaru) and slightly leached or non-leached Ferruginous Soils, in which there is little or no accumulation of clay with depth. Preliminary results in Nigeria tend to confirm this hypothesis, although adsorption may occur in non-leached soils if the pH is lowered sufficiently (Bromfield, personal communication, 1972). There is a relatively large leaching loss of 5—7 kg S/ha/annum from the sandy, slightly leached Ferruginous Soil in the Bambey lysimeters (Charreau, 1972), which is consistent with low sulphate adsorption.

(b) *Availability*

Sulphur deficiency occurs widely in the savanna and has been observed particularly in crops of cotton (Braud, 1967; 1969-70) and groundnuts

TABLE 35

Sulphur fractions in eleven soils (0—15 cm) in northern Nigeria, sampled in first year after bush fallow (Cooper, 1971).

		Range	*Mean*
Total nutrients:	Carbon %	0·17—0·83	0·41
	Nitrogen %	0·020—0·062	0·036
	Sulphur, mg/kg	18—54	35
	C:N:S ratio		114:10:0·99
*Sulphur fractions:	Organic:—		
	HI-reduceable	9—23	16
	Carbon bonded	2—10	6
	Inorganic:—		
	Adsorbed + soluble	2—8	4

*Accounting for 74 per cent of total soil sulphur.

TABLE 36

Total and residual sulphur in the plots of a 19-year-old fertilizer experiment at Samaru, Nigeria (Bromfield, 1972).

Treatment no.	*Annual application to soil (kg/ha)*	*kg S/ha (0—180 cm depth)*			
		Total in soil	*Residual (by diff.)*	*Applied in 19 yrs.*	*Residual/applied*
1	Nil	389	—	0	—
2	5000 FYM	613	224	?	—
3	112 AS	776	387	511	76%
4	112 AS + 112 SS	1102	713	788	90%
5	112 AS + 112 SS + 5000 FYM	1359	970	788+?	—

FYM = Farmyard manure; AS = Ammonium Sulphate; SS = Single superphosphate

(Greenwood, 1951; 1954; Nye, 1952a; Stephens, 1960b; Goldsworthy and Heathcote, 1963; 1964; Bolle-Jones, 1964; Bockelee-Morvan and Martin, 1966a, b; Lienart and Nabos, 1967; Bouyer, 1969). It does not appear to be confined to any one soil type (Vaille, 1970).

In the absence of large-scale industry the atmospheric sulphur contribution is small throughout the region, and except where fertilizers containing sulphur are used, or where previous land management has allowed a reserve of adsorbed sulphate to build up, mineralization of soil organic sulphur plays an important part in the sulphur nutrition of crops. Nye and Greenland (1960) stated that sulphur deficiency in West Africa was most acute after a high-grass (*Andropogoneae*) fallow, and results from Ivory Coast and Central African Republic on cotton (Braud, 1969-70) showed that deficiency was greatest in the first year after clearing from bush and thereafter declined. In Nigeria, crop response to applied sulphur correlated with soil C:S ratio (Goldsworthy and Heathcote, 1963), suggesting an analogy with the high soil C:N ratios and low nitrogen availability that occur after long grass fallows. Factors influencing the mineralization of organic sulphur have been discussed, in an Australian context, by Freney (1967).

Although C:S ratios may be a guide to probable mineralization rates, immediately available sulphur is better measured as the adsorbed + soluble sulphate extracted with 0·01 M monocalcium phosphate (Barrow, 1967). In a pot experiment with 15 Nigerian savanna soils Cooper (1971) found a significant correlation ($P = 0 \cdot 01$) between the dry matter yield of millet and the drop in adsorbed and soluble sulphate that occurred during crop growth.

However, the value of this fraction as a measure of sulphur availability to field crops is limited, in as far as it takes no account of the extent and rate of turnover of sulphur between organic and inorganic fractions during the growing season, nor of the subsoil reserves of adsorbed sulphate where

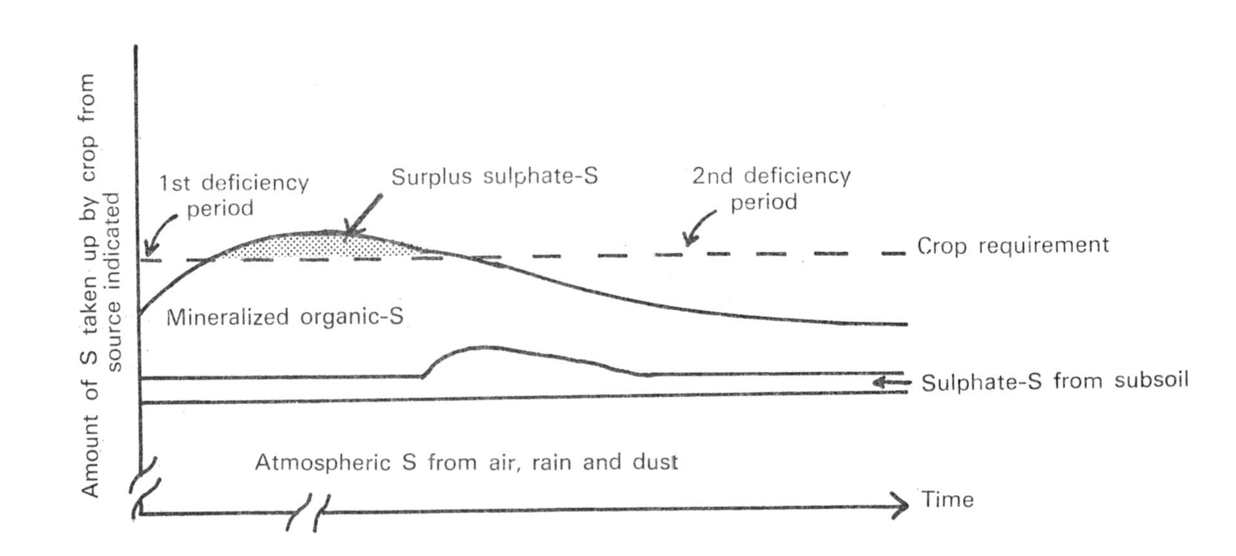

Fig. 7. Diagrammatic representation of the changes of sources of S to crops, from clearance of natural fallow through the cropping cycle (not to scale) where no S is supplied (Bromfield, 1972).

these exist. As Bromfield's work has shown, these subsoil reserves can be quite large, although their availability depends on the depth and on the rooting pattern of the crop. Deep rooting may often be discouraged by the compact nature of the subsoil, and shallow-rooted crops such as groundnuts, particularly if grown on ridges, may be unable to exploit subsoil sulphate. This may account in part for why sulphur deficiency has been observed most frequently in groundnuts (Bromfield, 1972).

The occurrence of deficiency in relation to natural sources of sulphur is illustrated by Fig. 7, taken from Bromfield (1972). Deficiency may occur when the fallow is first cleared, before there has been much mineralization of organic sulphur, and also after several years cropping; but in between there is a period when subsoil sulphate and mineralization may together meet the crop's requirements.

(c) *Gains and losses*

While a bush fallow may bring about some accumulation of mineralizable sulphur in the crop rooting zone despite losses when the vegetation is burned, ultimately the only source of sulphur for the soil apart from fertilizers is the atmosphere. Bromfield (1974a; 1974b) found that the annual deposition of sulphur at 11 sites along a 530 km north-east to south-west transect of the Nigerian savanna amounted to only 0.099—0·429 kg/ha (mean: 0·226 kg/ha) in the dust and 0·49—1·89 kg/ha (mean: 1·14 kg/ha) in the rain.

Losses through leaching and erosion depend on soil and site. That of 5—7 kg/ha/annum from the Bambey lysimeters (Charreau, 1972) strongly suggests an imbalance in these soils under prolonged cropping; but even where leaching losses are negligible it is doubtful whether an annual accession of 2 kg/ha or less is sufficient to sustain high crop yields. Bromfield (1974b) calculated that the mean annual atmospheric accession along his sampling transect was equal to the sulphur removed by 650 kg/ha of groundnuts (kernels + shells), which is about the yield obtained by local farmers not using fertilizers. A high yielding groundnut crop can take up more than 10 kg S/ha (Bromfield, 1973).

There is little information on sulphur uptake by other savanna crops, but using data of Whitehead (1964) one may calculate values of about 2—5 kg S/ha for moderate yields of cereals and cotton. In the case of cereals and cotton, but not groundnuts, some of the sulphur taken up may be returned to the soil in crop residues, but amounts recycled in this way are likely to be small, and it seems improbable that high yields will be maintained under savanna conditions without some external source of sulphur, as manure or fertilizer. Responses to fertilizer are discussed in Chapter 10.

2 POTASSIUM

(a) *Amounts*

Soil potassium content varies widely according to parent material, but compared with other nutrients amounts are often high (Table 37). In northern Nigeria, surface soils (0—15 cm) on basement complex rocks contained a

TABLE 37

Total potassium in savanna topsoils (0—15 cm)

Source	Country	Number of samples	Per cent K Range	Per cent K Mean
1	Ghana	5	0·14—1·05	
2	Ghana	11	0·11—1·02	0·38
3	Upper Volta	13	0·33—3·20	0·86
4	Nigeria	33	0·06—3·99	1·43

Sources: 1. Nye and Stephens, 1962;
2. Acquaye *et al.*, 1967;
3. Dupont de Dinechin, 1967c; 4. Wild, 1971.

mean of 2·2 per cent potassium, but mean content over sandstone was only 0·08 per cent (Wild, 1971).

Most of the potassium is associated with the sand and silt fractions (Acquaye *et al.*, 1967; Wild, 1971), where it occurs in orthoclase felspar and micas. In the clay fraction kaolinite is usually dominant and illite occurs only in relatively small amounts. Wild recorded a mean content of 0·90 per cent K for the clay fraction of his Nigerian soils, indicating an illite content of 10—15 per cent.

Amounts of exchangeable potassium tend to be low. A mean value of 0·25 me/100 g (range: trace—0·70 me/100 g) was given for 41 soils from Upper Volta (Dupont de Dinechin, 1967c). In the northern Nigerian soils referred to above the mean was 0·18 me/100 g, and the mean ratio of total to exchangeable potassium was 200. Nye and Greenland (1960) reported total to exchangeable ratios of 20—120 in five savanna topsoils (probably from the tall grass savanna) and 2·2—10·4 in soils from the forest zone, a difference which shows the greater content of weatherable minerals under dry savanna conditions.

(b) *Availability*

The availability of potassium in 33 soils from northern Nigeria was measured by Wild (1971) in terms of quantity/intensity relationships. In the upland soils (mostly Ferruginous and Ferrallitic) it was found that the intensity of potassium, expressed either as the ratio $K:\sqrt{Ca}$ or as K concentration (moles/l solution), was high, whereas the quantity of potassium measured by exchange with ammonium acetate or by linear extrapolation of the Q/I curve was low, and the buffer capacities for potassium were generally very low indeed. Hence, under intensive cropping most of the soils would rapidly become depleted of readily available potassium, a conclusion that was supported by the results of exhaustive cropping in pots. The soil was rapidly depleted of exchangeable potassium and there was only a small uptake of non-exchangeable potassium. Quantity parameters gave the best correlation with potassium uptake, and of these exchangeable potassium extracted with ammonium acetate was the easiest to measure.

A different pattern of behaviour is found in soil types with medium to high exchange capacity (e.g. Vertisols, Eutrophic Soils, and some Hydromorphic Soils). Although exchangeable potassium values are fairly high (0·1—1·0 me/100g), the proportion of potassium among the exchangeable bases is generally low. Under these circumstances one may expect high values for quantity parameters and for buffer capacity but quite low intensities.

This pattern is well illustrated in Table 38 with the data of Acquaye *et al.* (1967). Intensity was low in vertisolic, alluvial and forest soils, which had low potassium saturation of the exchange complex, but relatively high in upland (probably Ferruginous) savanna soils, in which potassium saturation was much higher. On the other hand, buffer capacity, being related to the size of the exchange complex, was greatest in the vertisolic soils and least in upland savanna soils. Under intense cultivation these differences between soil types will demand different patterns of potassium fertilization.

Measurement of the quantity/intensity relationship is too elaborate for routine use. However, potassium intensity is related to the degree of potassium saturation of the exchange complex, and buffer capacity to the cation exchange capacity, and for many practical purposes these two simpler measures may be used instead. Boyer (1972) adopted this approach for his concepts of absolute and relative minimum requirements to avoid deficiency in humid tropical soils: the exchangeable potassium in the soil must exceed both a certain absolute minimum level (between 0·07 and 0·20 me/100 g) and also a certain relative minimum level (at least 2 per cent of the sum of all exchangeable bases). These requirements are only approximate and will vary with the crop and the required level of production.

Another factor is the rate of release of non-exchangeable potassium. From soils with a kaolinite-dominated clay fraction such release is likely to be small. Wild's (1971) pot tests showed that there was some uptake of non-exchangeable potassium from all the savanna soils tested, although release was generally less than that normally found in temperate soils. A partial restoration of the exchangeable potassium occurred when the cropped soils were subjected to wetting and drying cycles. In a wider range of soils uptake of non-exchangeable potassium in a pot experiment varied considerably with soil type and clay mineralogy (Acquaye *et al.*, 1967). Release from non-exchangeable form was greatest in vertisolic soils and least in upland savanna soils.

Release from primary minerals may be important in some soils, but will depend on the degree of weathering of the profile and the nature of the parent material. Shallow soils over basic parent materials may be very poor in potassium if the parent rock lacks potassium-bearing minerals (Martin *et al.*, 1966; Kaloga, 1970).

(c) *Gains and losses*

Although potassium deficiency has not been widely reported from savanna areas, by the criteria discussed above many of the soils are potentially

TABLE 38

Potassium availability parameters in relation to soil properties (Acquaye *et al.*, 1967)

Soil type	*General soil properties*			*Parameters of soil potassium*			
	Clay, %	*Carbon,* %	*C.E.C., me/100 g*	*K saturation of soil C.E.C., %*	*Intensity, AR X* 10^2, $(M/l)^{\frac{1}{2}}$	*Quantity*	*Buffer capacity, me/100 g* / $(M/l)^{\frac{1}{2}}$
Vertisolic	30	1·40	23·7	0·58	0·3	0·18	51·4
	26	1·37	32·0	0·54	0·3	0·26	86·7
	25	1·11	24·3	0·70	0·2	1·52	76·0
Hydromorphic alluvial	46	2·10	21·3	1·49	0·6	0·25	41·7
Ferrallitic	43	3·56	23·8	0·95	1·0	0·24	24·0
	27	1·40	12·0	0·56	0·6	0·09	15·0
	21	1·85	13·5	0·89	1·0	0·13	13·0
*Ferruginous	13	0·59	4·2	3·28	2·4	0·16	6·7
	33	1·56	17·7	1·97	1·3	0·29	22·3
	12	0·28	3·1	3·18	1·9	0·12	6·3
**Ferruginous (hydromorphic)	11	0·26	4·2	2·66	1·4	0·15	10·7
	48	1·08	13·1	1·94	0·8	0·17	21·2
	16	0·42	3·5	4·74	3·8	0·18	4·7
	8	0·42	3·3	3·92	4·4	0·13	2·9

*Savanna ochrosols; ** Ground-water laterites.

deficient, and this deficiency seems certain to show itself when yields are raised. It is therefore important to consider the long-term balance of available potassium.

Leaching losses at Bambey, Senegal are of the order of 3—12 kg K/ha/annum (Vidal and Fauche, 1962; Tourte *et al.*, 1964). At the higher end of this range losses may be of agricultural significance; 12 kg/ha represents 15 per cent of the exchangeable potassium in a topsoil (0—15 cm) having 0·10 me/100 g exchangeable potassium. Leaching losses of calcium and magnesium are usually much greater, but so also are the exchangeable reserves of these ions. In the *in vitro* leaching studies of Moureaux and Fauck (1967) the equivalent of 4000 mm of rain reduced exchangeable potassium in a Brown Sub-Arid, a Ferruginous and a slightly Ferrallitic Soil to 55, 35 and 35 per cent of their respective original values; corresponding figures for exchangeable calcium were 99, 78 and 83 per cent.

Moreover, there is similar evidence from the field. Reports of changes in exchangeable cations under cropping show the greatest proportional decline to be in exchangeable potassium. During 15 years cultivation at Sefa, Senegal, exchangeable potassium dropped from 0·10—0·20 me/100 g to 0·01—0·05 me/100 g (Charreau and Fauck, 1970), a loss of 100—200 kg K/ha, attributed to leaching and erosion. Over the same period exchangeable calcium dropped from 2—3 me/100 g to about 1 me/100 g. Hence, there was both a relative and an absolute decline in exchangeable potassium, i.e. a decline in both quantity and intensity parameters. In these soils the loss over 15 years had apparently been borne entirely by the exchangeable fraction, for there had been no decline in the non-exchangeable potassium reserves. Similar results have been obtained in the groundnut-growing areas of Senegal (Prevot and Ollagnier, 1959). It may be that in these soils, whose reserves of potassium occur mainly in the silt and sand fraction, release into exchangeable form is so slow as to be of only minor agricultural significance.

A further aspect of the delicate balance of soil potassium was shown in Nye's (1958a) discussion of the accumulation of nutrients by fallow vegetation. Although amounts of exchangeable calcium and magnesium in the top 25 cm of soil greatly exceeded the amounts of these ions in the vegetation, the amount of potassium held by the grass and trees of natural savanna woodland was about equal to that in exchangeable form in the soil. Hence the importance to subsequent crops of the potassium released by fallow burning; though whether repeated burning causes a significant loss of potassium appears not to be known.

Most important under continuous agriculture is crop removal; and it is particularly necessary to distinguish here between total uptake of potassium and actual removal in the economic yield (Velly, 1972). For example, to produce 6—8 tonnes of maize dry matter per ha, total uptake of potassium is in the range 100—200 kg K/ha, but the potassium in the grain accounts for only 25—30 kg of this. The long-term balance, and the size of the eventual fertilizer requirement, therefore depends very much on how the crop residue is disposed of. In general, farmers' present practices tend to conserve potassium, because residues are removed from the field only when they have practical or economic value. Groundnut haulms are carried off for fodder, but much of the cereal residue is left, burned or grazed *in situ*. However, the practice in field experiments of removing all residues from the field, for practical convenience, may induce potassium deficiency where it would not otherwise exist and lead to unnecessary recommendations for potassium fertilizer being made to farmers.

3 CALCIUM AND MAGNESIUM

(a) *Amounts*

Calcium and magnesium are the dominant exchangeable cations in virtually all savanna soils, and few deficiencies have been reported. Topsoil contents vary widely, but, except in Vertisols and Eutrophic Brown Soils which may have higher values, most fall within the limits 2—20 me total Ca and 1—20 me total Mg per 100g soil. Between 25 and 80 per cent of the total calcium and 20 and 60 per cent of the total magnesium usually occur in exchangeable form. In this respect both ions differ markedly from potassium, and loss from the soil on cultivation may represent a high proportion of the total reserves.

In most topsoils exchangeable calcium exceeds exchangeable magnesium, but a wide range of exchangeable Ca:Mg ratios has been reported: between 1:1 and 7:1 in "sols rouge" in north Cameroun (Martin *et al.*, 1966); between 2:1 and 4:1 in Ghana, being lower in less well-drained soils (FAO, 1967); and between 2:1 and 10:1 in the sub-arid soils of Senegal (Maignien, 1959). These ratios commonly decrease with depth.

Estimates of leaching losses of calcium plus magnesium at Bambey ranged from 20 to 79 kg/ha/annum, the calcium loss being rather greater than that of magnesium (Table 39). Similarly, at Bambari, Central African Republic (Quantin, 1965), loss of exchangeable calcium from cultivated land was relatively greater than that of magnesium; but at Niari, Congo (Martin, 1970) magnesium was lost more rapidly than calcium. One consequence of these losses may be a change in the ratio of cations on the exchange complex. During 15 years cultivation at Sefa greater proportionate losses of calcium over magnesium narrowed the exchangeable Ca:Mg ratio from 2—3 to about 1 (Fauck *et al.*, 1969).

The *in vitro* leaching studies of Moureaux and Fauck (1967) implied that there might be large differences between soil types in relative rates of cation leaching (Table 39), but to what extent these results may be generalized is not clear. Referring to the changes at Sefa, Fauck *et al.* (1969) commented on the importance of a good exchangeable Ca:Mg ratio, particularly for structure, and suggested the value of 1·0 as the lowest acceptable limit.

There is little information about rates of release from non-exchangeable forms, but Charreau and Fauck (1970) noted that losses of calcium from Sefa soils over 15 years had, like those of the potassium, been entirely from the exchangeable fraction, and the non-exchangeable calcium reserves had not been encroached on.

(b) *Availability*

Little work has been done on measuring the availability of calcium and magnesium or in assessing the adequacy of amounts, when these are low, for crop nutrition. However, deficiency of calcium in groundnuts causing low shelling percentages has been reported from some sandy soils in northern Nigeria (Anon, 1959; Meredith, 1965). Blind nuts were common when soil exchangeable calcium was 0·04—0·08 me/100 g, whereas in comparable areas producing full nuts it was 0·15—0·24 me/100 g. Cotton is also known to be a crop with a high calcium requirement, but no responses attributable to calcium have been reported from the savanna.

More widespread deficiencies may occur under intensive cropping where fertilizer use is restricted to the application of N, P, K and S. The level of exchangeable calcium and magnesium may fall because of repeated removal of the nutrients in crop produce and because of leaching losses. Also, adverse nutrient ratios, for example a high K:Mg ratio, may be caused by fertilizer use. Hence, in soils with low cation exchange capacities a carefully calculated addition of calcium and magnesium may prove to be required either for sensitive crops or regularly once per rotation.

As with potassium, the return of crop residues to the soils, either directly or as ash, reduces loss, for very little of the calcium or magnesium taken up by the crop is removed in the economic yield (Bertrand *et al.*, 1972). However, the frequent return of ash to the topsoil may, like an unbalanced fertilizer dressing, lead to an imbalance of exchangeable cations through a relative concentration of potassium derived from deeper horizons.

TABLE 39

In vitro leaching losses of exchangeable calcium and magnesium from soils of different types (Moureaux and Fauck, 1967).

Soil type	*Exchangeable cations, me/100 g*			
	Originally in soil		**Lost on leaching*	
	Ca	*Mg*	*Ca*	*Mg*
Vertisol	33·91	6·58	0·00	0·75
Brown Sub-Arid	8·14	2·11	0·05	0·60
Ferruginous	1·84	0·59	0·39	0·12
Slightly Ferrallitic	3·08	1·49	0·51	0·39

* 4000 mm simulated rainfall.

TABLE 40

Mean trace element contents of some savanna soils.

Source	*Soil type*	*Number of samples*	*Country*	*Trace elements in the soil, ppm*					
				B	*Mn*	*Cu*	*Co*	*Zn*	*Mo*
1	Ferruginous		Dahomey	—	155	<1	2	65	—
2	Savanna (probably slightly leached Ferruginous)	7	Mali	0·2	229	22	4	3	0·7
3	Brown soils	8	Cameroun	2·7	2165	68	52	—	—
	Ferrallitic	47		4·8	576	39	15	—	—
	Ferruginous	28		5·4	546	18	13	—	—
	Sodic	10		6·6	351	15	11	—	—
	Vertisols	8		9·2	1086	37	26	—	—
	Hydromorphic	9		3·4	111	12	7	—	—

Sources: 1. Pinta and Ollat (1961), by arc emission spectrography.
2. Peyve (1963), analysis "after treatment with acids"; no further details given.
3. Nalovic and Pinta (1972), by arc emission spectrography.

4 MICRONUTRIENTS

(a) *Amounts*

Information on trace elements (micronutrients) in savanna soils is very limited. A summary of the only analyses found in the literature is given in Table 40, and this shows a wide variation in values. The most comprehensive study was that of Nalovic and Pinta (1972) in Cameroun, and their main conclusions can probably be applied to the region as a whole.

These conclusions were that the nature of the parent material determines the trace element content of the horizons of weathering (BC and C), but that pedogenetic factors are more important in the upper horizons (A and B). In general, soils over crystalline and metamorphic rocks give rise to soils richer in trace elements than those over sedimentary rocks; but differences were noted between different soil types formed on similar parent material. Thus Vertisols, red Ferrallitic Soils and certain Ferruginous Soils were rich relative to Hydromorphic and yellow Ferrallitic Soils. The richest soil was a young brown soil formed over basalt.

The importance of pedogenetic factors was greater for the alkaline and alkaline earth elements (Sr, Ba, Li, Rb) and least for the heavy metals (Mn, Pb, Ga, Cr, V, Cu, Ni, Co). The latter are less mobile and tend to be absorbed by the iron oxides; for example, a positive relationship between soil heavy metal content and the amount of amorphous oxides was noted. No correlation was found between trace element content and either soil pH or organic matter content. In the case of organic matter, the lack of any relationship was attributed to the rapidity of organic matter decomposition under tropical conditions and the consequent failure to evolve molecules suitable for the formation of organometallic complexes.

(b) *Availability*

The great age and highly weathered state of many savanna soils leads one to expect the availability of most micronutrients to be low; yet there have been relatively few reports of deficiencies. An early summary was given by Schütte (1954). Various reasons may be suggested: (i) under subsistence farming yields are low and nutrient removal is low, (ii) nutrients in the crop are largely returned in ashes, animal dung, domestic refuse etc., (iii) the fallow period effects a build-up in the topsoil similar to that of the major nutrients, and (iv) deficiencies have been overlooked, especially where yields have been limited by other factors. Thus, in northern Nigeria, symptoms of boron deficiency in cotton, first recognized only in 1969 on experimental plots (Faulkner, 1971), have since been found to occur commonly on farmers' crops (Heathcote and Smithson, 1974). Also, Webb (1954) pointed out that the presence of several mineral deficiencies simultaneously may lead to failure to identify one of them. In pot experiments with Gambian soils he showed that severe deficiencies of nitrogen, phosphorus and molybdenum could all occur in the same soil. Of the other nutrients, copper, zinc and probably boron also showed some signs of deficiency.

The two micronutrients most commonly deficient are boron in cotton (Braud *et al.*, 1969; Fritz, 1971; Heathcote and Smithson, 1974), and molybdenum in groundnuts (Stephens, 1959; Martin and Fourrier, 1965; Gillier, 1966; Heathcote, 1972a). Responses to each of these elements have been obtained.

Boron occurs in organic matter, in the soil solution and as weakly adsorbed forms on clay surfaces. The generally low amounts of organic matter and clay in these soils means that the danger of deficiency is widespread and that crops are very dependent on boron in the soil solution. This perhaps accounts for the observation by Heathcote (1972a) that boron deficiency is less acute in soils which stay moist longer. Deficiencies may also be induced by high rates of leaching, where these occur.

Deficiency of molybdenum would be expected to be most severe on acidic soils high in active iron oxides. Hence a deficiency might result from the use of ammonium fertilizers, and be rectified by liming. According to Gillier (1966), molybdenum deficiency in groundnuts is common in the north of Senegal, but the effect of applied molybdenum on pod yield has been erratic because it also increases leaf area and therefore drought susceptibility. More consistently, molybdenum has increased the yield and nitrogen content of the haulm (Gillier, 1966; Heathcote, 1972a), which is itself useful because of the value of the haulm as fodder. Acute deficiency has not yet been reported.

The frequency and severity of occurrence of these and other micronutrient deficiencies may be expected to increase as the intensity of agriculture increases. Higher yields, more frequent cropping, and the export of produce seem bound to overtax the resources of many soils, as will the shortening of fallows and the replacement of organic manures and other domestic wastes by inorganic fertilizers. The current change from older types of inorganic fertilizers to higher analysis materials containing less micronutrient impurity (Sauchelli, 1969) will further aggravate the situation. Hence supplementary micronutrients, supplied either as additives to major nutrient fertilizers or as sprays, will probably become a requirement in some areas. Care will then be needed to avoid adding harmful amounts, especially of boron, because most soils are low in clay and organic matter, the two constituents that moderate toxic effects.

Chapter 9

EFFECTS OF TRADITIONAL PRACTICES ON SOIL FERTILITY

As we have written earlier, the agricultural system throughout the savanna has been, and still largely remains, one of subsistence farming. However, increasing populations, and particularly increasing urban populations, will create the need for higher productivity and for the production of food cash-crops for a consumer market. At the present time it seems likely that these needs will be met largely by increases in production by small farmers, through a gradual adaptation and intensification of the traditional system. Such changes have already begun in more densely populated areas. In this chapter we review the relationships between traditional agricultural practices and the maintenance of soil fertility and attempt to gauge the impact of agricultural change upon them.

1 FALLOWING

The most important feature of the traditional system is the deliberate abandonment of farmland, after a variable period of cultivation, to a naturally regenerated fallow. The nature of the fallow vegetation varies according to the climatic zone, the age of the fallow, the frequency of fire, and to such site factors as soil depth, drainage and fertility. In the southern savanna the regrowth of coppiced shrubs and trees quickly takes over from the grasses and shrubs that initially succeed the crop plants. In the north the fallow consists of a succession of grass species (Morel and Quantin, 1964), and woody plants invade only slowly, if at all. However, a large part of the savanna is an intergrade between these two extremes, with both grasses and trees playing some part in the fallow. Before a new phase of cultivation is started, the fallow vegetation is completely or partially cut down and burned.

The main effects of the fallow (Nye, 1958a; Vine, 1968) are:

(i) to accumulate and store nutrients in the vegetation for eventual release to the topsoil (by litter-fall, leaf wash, root decomposition, and through the ash of the final burn);

(ii) to increase the amount of organic matter in the topsoil and thereby its total nutrient content and cation exchange capacity;

(iii) to suppress the weeds, pests and diseases of cultivation.

The relative importance of these functions depends on local conditions. Although we may assume that the fallow always effects some build-up of fertility, land is not necessarily abandoned to bush only when fertility is exhausted. In some areas it is often easier to clear new land from bush than to continue to fight against vigorous weed growth (Nye, 1958b; Vine, 1968; Jurion and Henry, 1969).

(a) *Nutrient accumulation*

The most important aspect of nutrient storage by the fallow is its accumulative nature. Through the extensive and variegated rooting systems of its component species, the fallow acts as a sink, intercepting and retaining a large proportion of all available nutrients. Moreover, uptake is from a greater depth of soil than is usually tapped by the roots of an annual crop, particularly if there is time for deep-rooted trees to develop. Some losses occur through fires, mineralization and leaching, but there is also much recycling. Thus when the fallow is eventually cleared and burned, the subsequent release of accumulated nutrients represents the greater part of all the nutrients that became available during the period of the fallow.

The relative importance of this effect varies for individual nutrients according to local conditions. In the extreme south under rainforest, where the soils are often highly leached and deeply and intensely weathered, the natural fertility of the mineral soil is generally low and a large proportion of the nutrients in the soil-plant system, particularly the cations, is held in the vegetation and the litter layer. On the Ferrallitic Soils of the well-wooded southern savanna the situation is similar if less extreme; but much of the central savanna is covered by Ferruginous Soils which are less leached and which often have appreciable reserves of weatherable minerals (d'Hoore, 1964). Since the natural vegetation here is much less dense than that of the rainforest zone and surface accumulations of litter are lighter, it may be supposed that clearing and burning releases a smaller quantity of plant nutrients. However, because of the poverty of these soils in available phosphate, even this small release may be essential to the successful production of a crop.

The only extensive data are those of Nye (1958a), who analysed fallow vegetation at the three stages of the succession, *Imperata cylindrica*, *Andropogon gayanus*, and savanna woodland with climax grasses, at a site near the southern boundary of the savanna (Ejura, Ghana). The grasses attained their highest nutrient storage (19, 83, 58 and 33 kg/ha of P, K, Ca and Mg, respectively) in the first two or three years of the sub climax fallow, *Andropogon gayanus*. Thereafter, all further storage was attained by the growth of woody vegetation, which in well-wooded savanna stored as much phosphorus and very much more potassium, calcium and magnesium than the grasses. Soil analyses showed that amounts of exchangeable potassium in the top 30 cm of soil were low, but amounts of exchangeable calcium and magnesium greatly exceeded those held in the vegetation. Nye suggested that the fallow plays an important part in conserving potassium but is less important for calcium and magnesium.

Over much of the savanna the most important nutrient released when a fallow is cleared is phosphate. In Ghana, crop responses to applied phosphate decreased with increasing length of previous fallow (Nye, 1952a; 1954; Stephens, 1960a), and it was estimated that a good fallow releases, on clearing, phosphate equivalent to 120 kg single superphosphate per ha (Nye, 1958a). However, the effectiveness of the fallow depends very much on the initial state of the soil and its influence on the nature and rate of establishment of the fallow vegetation. Stephens (1960c) reported that grass fallows at Ejura benefitted succeeding crops through raising soil phosphate status, but on a poor soil at Nyankpala the growth of grass was rarely sufficient to make any effective impact on soil phosphate status. One may conclude that fallowing alone is unlikely to bring severely degraded land back into agricultural production.

Nye and Foster (1961) investigated the source of the phosphate accumulated by a high grass savanna. They found that perennial grasses and the associated dicotyledons derived over 30 per cent of their phosphate from below 25 cm. Since a similar fallow had been shown to add 6·7—13·5 kg P/ha to the surface soil every year (Nye, 1958a), it follows that under this cover some 2·2—4·4 kg P/ha are brought up from below 25 cm annually.

Most of the nitrogen held in the fallow vegetation is lost on burning, but that in the soil organic matter is retained. Although this accumulation is presumed to derive largely from microbiological fixation, two other mechanisms have been suggested. Jones (1968), working on Ferrallitic Soils in Uganda, found that much of the nitrogen gained by the plant-topsoil system during the fallow period derived from nitrate leached into the subsoil during the preceding arable period. The widespread occurrence in the savanna of soils having a textural B horizon with a small nitrate sorption capacity (see Chapter 6) might be expected to promote a similar cycling mechanism.

There is also the suggestion of Jaiyebo (1967), based on data of Moore and Ayeke (1965), that non-exchangeable ammonium-N in the subsoil may contribute to nitrogen accumulation in the fallow. The four profiles studied by Moore and Ayeke, three in the Nigerian forest zone and one in the Sahel zone, contained between 1320 and 2390 kg ammonium-N per ha, extractable with hydrofluoric acid, in the top 155 cm. As these authors pointed out, the release of only one per cent of this fixed ammonium would make available 13 to 24 kg N/ha, enough, they suggested, to make a significant contribution to the biological nitrogen cycle, especially as this source is not lost on burning.

Even quite short herbaceous fallows may benefit the potassium nutrition of subsequent crops. In the potassium-deficient Patar area of Senegal groundnut yields were increased by 385 and 570 kg/ha, respectively, by two and three-year fallows relative to a one-year fallow; and negative interactions between fallow length and response to potassium fertilizer have been reported from both Senegal and Gambia (Bockelee-Morvan, 1964; Evelyn and Thornton, 1964; Gillier and Gautreau, 1971).

(b) *Effect on soil physical properties*

Morel and Quantin (1964) followed the successive stages of flora establishment in natural fallows in the Sudan zone and correlated floristic composition with the structural status of the soil. They found that each of the three or four principle floristic stages of the fallow corresponded to a certain structural status of the soil, that the rapidity with which these stages appeared depended upon the initial state of the soil, and that the erect, deeply-rooted grasses forming the third stage of the succession significantly and rapidly improved soil structure.

Other workers have reported improved structure after grass fallows (Martin, 1944; Pereira *et al.*, 1954; Stephens, 1967; Wilkinson, 1970), but the rate at which this improvement is achieved seems to be much slower than that of the structural degradation that occurs under cropping (Combeau and Quantin, 1963). All benefit may have been lost after one year of cultivation (Pereira *et al.*, 1954; Wilkinson, 1970). Soil texture and fallow species may be factors in determining the persistence of improved physical properties. Thus in Uganda, Stephens (1967) found that structural improvements in a clay loam after elephant grass (*Pennisetum purpureum*) survived two years of cropping.

Such physical effects are caused mainly by root proliferation. However, an additional benefit may be achieved by ploughing into the soil the residues of an herbaceous fallow instead of burning them (Charreau and Fauck, 1970). This practice, which is effectively a green-manuring operation, will be discussed more fully in Chapters 10 and 11.

(c) *Effect of length of fallow*

Both the physical and chemical changes brought about by the fallow are directly dependent on its length and composition. The accumulation of soil organic matter and of mineral nutrients are gradual processes. A short fallow does not allow time for deep-rooted species, particularly trees, to develop, and on the deeply leached soils of the forest zone and parts of the moist savanna the presence of deep-rooted species is particularly important if the resources of the whole profile are to be fully exploited.

Nye's data for Ejura (Nye, 1958a) showed that half the accumulated nutrients in mature savanna woodland were in the woody vegetation. Vine (1968) also emphasized the importance of trees; and the widespread indigenous practice of preserving the rootstocks of shrubs and trees of the previous fallow to facilitate the rapid establishment of the new fallow is evidence of the importance attached by the local cultivator to the presence of woody vegetation in the fallow.

Nevertheless, in the drier northern savanna, where tree cover is sparse or absent, fallowing under grass or grass-shrub mixtures is an effective means of restoring fertility. This may be because the relative advantage of deep-rooted species is reduced in soils that are less weathered and more base-saturated. Further, although the nutrient content of such fallows is usually very much less than that of forest fallows, the lower rainfall and lower levels of nitrate in these grassland soils (Stephens, 1960c; Djokoto and Stephens, 1961a) make cation loss by leaching a much less serious problem. The first limiting nutrient in these soils is phosphate, and there is evidence that this nutrient is well supplied to subsequent crops by good grassland fallows (Stephens, 1960c).

The effectiveness of a fallow in restoring the soil may depend as much upon the degree of its impoverishment at the end of the previous cropping period as upon the length of time the fallow is allowed to grow, for the establishment of effective fallow species may be much slower on badly degraded land (Morel and Quantin, 1964). Where land is in short supply the whole nature of the fallow may change. Watson and Goldsworthy (1964) reported that shortened fallows and the cutting of wood had caused the Guinea savanna bush vegetation in parts of the Nigerian 'middle belt' to be replaced by grass.

Estimates of the fallowing-cropping ratio necessary for the maintenance of fertility vary widely. Nye and Greenland (1960) stated that cropping periods of some two to four years followed by six to twelve years of fallow were common and apparently stable; and that although many examples of three years cropping following three years fallowing exist, at this intensity fertility falls to a very low level. Dennison (1959), working in Nigeria, made the point that where fallows are very short, although fertility declines, the decline in each cycle may be quite small and not be noticed for some time; but once the soil has become infertile, a much longer fallow is required to restore it. Porteres (1952) gave the minimum period of fallow as two to three times as long as the cropping period and up to fifty times as long where fertility is low. The existence of these very wide differences in ratios is a reflection of several factors: inherent soil fertility, the nature of the fallow vegetation, population pressure and, possibly, cultivation techniques. Where organic manures, ashes or inorganic fertilizers are applied a lower ratio can probably be tolerated.

The optimum length of fallow was investigated experimentally by Watson and Goldsworthy (1964) in Nigeria. They found that a minimum

TABLE 41

The effect of fallow length on subsequent crop yield in Nigeria (Watson and Goldsworthy, 1964).

Fallow length (years)	*Crop yield, kg/ha Year 1		Year 2	Year 3
	Yams	Cowpeas	Millet/sorghum	Benniseed
1—2	5660	116	966	336
3	9680	126	1196	346
4	10790	149	1080	252

*Means of crops grown in a (yam + cowpeas)—(millet + sorghum)—benniseed rotation following fallows of *Andropogon gayanus* and of *Cajanus cajan*.

of three years of fallow was necessary after every three years of cropping (Table 41), but any additional advantage from a longer fallow was almost entirely confined to crops grown in the first year after clearing. However, yields were only modest, and these authors considered that more continuous cropping with fertilizers held greater promise than did fallowing.

In other experiments in Nigeria cotton yields were higher after a six-year than after a two or three-year fallow of *Andropogon gayanus*, but the yield of the second crop, sorghum, was affected by previous fallow length at only one site, and that of the third crop, groundnuts, not at all (Goldsworthy and Meredith, 1959). Again, under much drier conditions at Louga, Senegal (Charreau and Tourte, 1967; Delbosc, 1967), natural fallow markedly increased groundnut yields, but the difference between a two-year and a six-year fallow was not significant (Table 42).

The apparent conclusion from this work is that long fallows give little extra benefit; but this may be misleading. The benefit referred to was measured solely in terms of crop yield in the years immediately following the fallow, whereas the greatest difference between long and short fallows is likely to be in their ability to maintain fertility over long periods. This has not been investigated experimentally.

(d) *The future of fallowing*

Bush fallowing is an inefficient form of land use where land is scarce and populations are increasing. In these circumstances the tendency is for fallows to become shorter and therefore less effective in maintaining fertility. To prevent complete soil degradation a number of possible courses are open:—

(i) to use planted fallows of species more efficient in restoring fertility than the naturally regenerating vegetation;

(ii) to use fallows, natural or planted, in combination with fertilizers so that the restoration of nutrient status is not totally dependent upon the action of a short fallow;

TABLE 42

The effect of fallow length on subsequent crop yield in Senegal (Delbosc, 1967).

Treatment	Groundnut yields, kg/ha Without fertilizer	With fertilizer*
Before fallowing	355	—
After 2 years fallow	713	913
After 3 years fallow	781	1039
After 6 years fallow	882	1075

*NPK fertilizer, amount not stated.

TABLE 43

Dry matter production of fallow vegetation in Senegal, t/ha. (Delbosc, 1967).

Fallow length (years)	Fallow treatment: Unburnt		Annually burnt	
	*Fertilized	Unfertilized	*Fertilized	Unfertilized
2	6·9	3·9	4·9	3·3
3	5·1	3·9	5·5	4·0
6	6·4	6·2	4·5	5·7

*NPK fertilizer applied in previous cropping cycle; amounts not stated.

(iii) to abandon the use of fallows completely and to rely on other means for the maintenance of fertility. This possibility is discussed in the following chapter.

Although many species have been tried, few planted fallows have proved significantly better than natural regeneration. In central Nigeria, pigeon pea (*Cajanus cajan*) and Gamba grass (*Andropogon gayanus*) were slightly more effective than natural regeneration where only a short fallow (four years or less) was possible, but such fallows were considered to have only limited value (Dennison, 1959; Watson and Goldsworthy, 1964). In a range of legumes tried in Central African Republic, almost all proved ineffective, even harmful, and Quantin (1965) concluded that tall grasses (e.g. *Panicum maximum, Pennisetum purpureum*) made the best fallows under humid conditions. It was suggested that the rooting habit is important. At Grimari, Central African Republic, a 6-year fallow of *Cajanus indicus*, a legume with a tap root, was comparable in effect with a fallow of *Pennisetum purpureum* of similar age (Morel and Quantin, 1972).

However, it is unlikely that a planted fallow will be acceptable to a farmer unless it has a demonstrable cash value. Perhaps the best returns would come from a herbage legume such as "stylo" (*Stylosanthes gracilis*) where the farmer keeps stock or where he can sell the crop to those that do.

Although fallows are themselves rarely fertilized, they may benefit considerably from the residual effects of fertilizers applied to previous crops (Schilling, 1965). In a trial in Senegal, Delbosc (1967) showed that fertilizers applied to crops had an important effect on dry matter accumulation in subsequent two and three-year fallows, but not where the fallow lasted six years (Table 43).

One may suppose that where short fallows are in rotation with fertilized crops the effect tends to be cumulative, leading to higher annual yields from fertilized rotations including short fallows than from unfertilized rotations including long fallows. This may explain the success of the "semi-intensive" system recommended in the groundnut areas of Senegal and other parts of francophone West Africa, in which a groundnut-cereal-groundnut rotation follows one or two years of natural fallow. Small amounts of NPK fertilizer are applied to the crops, and the unburned vegetation is incorporated into the soil with animal traction at the end of the fallow. Under this system groundnut yields have been maintained at about 2000 kg/ha for 15 years (Delbosc, 1967).

2 BURNING

Fire is one of the most powerful tools in the hands of the savanna farmer. Three different spheres of use may be defined:—

(a) *Bush fires*

Bush fires may be started to stimulate growth of fresh grazing or to flush out game for hunters, or they may be started accidentally. Because of their uncontrolled progress and the indiscriminate destruction that they cause, they have been the subject of much controversy, legislation and scientific study (see bibliography of Bartlett, 1957, and reviews of Guilloteau, 1956, and Ramsay and Innes, 1963); and they have frequently been blamed for deteriorating fertility (e.g. Ferguson, 1944; Jewitt, 1949; Bachelier *et al.*, 1956). However, the intensity of a burn and its destructive power depend on the season. A fire early in the dry season does much less damage than one at the end, and an early burn may be used, for example by foresters, to forestall the possibility of a destructive late burn.

Where late season fires occur annually their effect on the vegetation is profound. All but the most fire-tolerant trees are eliminated, and the general effect is to encourage grasses at the expense of trees (Trapnell, 1959; Keay, 1960; Ramsay and Innes, 1963 ; Rains, 1963 ; West, 1965). Hopkins (1965b) reported that in the derived savanna of western Nigeria late burning for five consecutive years reduced the tree population by 32 per cent, and regeneration was virtually prevented. Some authorities regard the vegetation of large parts of the savanna as a fire climax (Aubreville, 1949).

There seems to be little doubt that frequent late-season bush fires are detrimental to the effectiveness of the bush as a fallow and to soil fertility. The following effects may be involved:

(i) Frequent fires tend to replace wooded fallows with grass fallows, which are less effective in accumulating and storing plant nutrients and in building up soil organic matter (Charter, 1955; Nye, 1958a; Nye and Greenland, 1960).

(ii) Burning destroys litter, thereby reducing the accumulation of organic matter in the surface soil. Hopkins (1965b; 1966) observed that "three-quarters of the annual production of the aerial shoots of herbs is lost to the biological system when the savanna is burnt" after the middle of the dry season, and that more leaf litter is destroyed by fire than by decay. In Kenya, Edwards (1942) measured a distinct increase in the humus of grassland protected from burning for ten years. A light early burn, however, may be beneficial. In the derived savanna of Nigeria, thirty years of annual burning raised the humus content of the upper 20 cm of the soil by 17 per cent where the burn was early, but lowered it by 12 per cent where the burn was late (Moore, 1960).

(iii) Burning, late or early, causes a loss of nitrogen, sulphur, and possibly phosphorus, from the vegetation and litter. At Bambey a burned fallow lost up to 20 kg N/ha in one year (Vidal and Fauche, 1961, 1962). On the other hand, Daubenmire (1968), in his review of the ecology of fire in grasslands, found that "whereas few investigators report significant loss of nitrogen from grass fires, and others find burning neither detrimental nor beneficial, many have found nitrogen conditions definitely improved in consequence of grass fires". He quoted the result of Moore (1960) that soil nitrogen was lowest in savanna plots that were protected from burning. This effect, it seems, may be due to increased growth of legumes after fire.

(iv) The destruction of litter and much standing vegetation exposes the soil to the violence of the early rains and increases the risk of erosion, although there is little evidence that severe erosion does occur at this time. Only a negligible loss of soil and dissolved solids was recorded from both burned and unburned bush fallow at Samaru, Nigeria (Kowal, 1970c) and at Sefa, Senegal (Roose, 1967). On the other hand, reports from southern Africa (Daubenmire, 1968) suggest that the increased rate of runoff after grass is burned can cause serious erosion even from relatively flat land.

(v) Burning has a selective effect on grass species and may tend to eliminate those most suitable for grazing (Brockington, 1961; Ramsay and Innes, 1963). In cultivated land, shortened fallows and frequent burning may encourage invasion by the rhizomatous grass, *Imperata cylindrica* (Watson and Goldsworthy, 1964).

(vi) The heating from a bush-fire may have a sterilizing effect on the surface soil. In Kenya bush burning was reported to have killed aerobic N-fixers and nitrifying bacteria, but not anaerobic N-fixers (Meiklejohn, 1955); but in Madagascar bush fires had a stimulating effect on nitrifying bacteria (Dommergues, 1954).

(b) *Bush clearing*

Bush clearing prior to a new cycle of cultivation necessarily includes the burning of the existing plant cover; for in traditional agriculture burning provides the only practicable means of disposing of the coarse clumps of grass left by a grass fallow and the large woody residues of a well-wooded fallow. At the same time it ensures that the mineral nutrients stored in the fallow are released in time for the first crop. After a grass fallow, fertility is limited by lack of available nitrogen, and this would probably be aggravated by incorporation into the soil of coarse grass residues (Nye, 1950; Stephens, 1960c; Djokoto and Stephens, 1961a). However, Singh (1961) reported that residues of two and three year grass fallows left on plots to decompose benefitted the subsequent crop indirectly by improving the physical condition of the soil, especially its structure and water regime.

The nutritional benefit of burning a fallow is illustrated by the results of Rains (1963) at Shika (Northern Guinea zone, Nigeria). Yields of the subsequent crop, sorghum, were greatly increased by burning a green manure compared with ploughing it in:—

Green manure	*Sorghum yield (kg/ha)*	
	Manure ploughed in	*Manure burned*
Crotolaria retusa	2057	2662
Mucuna (*Stizolobium deeringianum*)	1313	1979
Bush grass fallow	424	806

Similar results were obtained at Ibadan by Vine (1953), who grew maize after burning or incorporating mucuna; and Heathcote (1970) obtained significantly higher yields of sorghum, cotton and maize after burning, rather than incorporating, mature bush grass on experimental plots. In each case the effectiveness of burning was attributable to the greater resulting availability of phosphate.

Apparently contrary results have been obtained in Senegal. For instance, Charreau and Poulain (1964), working on sandy 'dior' soils, obtained higher millet yields when the preceding herbaceous fallow was incorporated than when it was burned:—

Fallow treatment	*Millet yield (kg/ha)*	
	No fertilizer	*NPK applied*
Burned	912	1565
Incorporated	1244	1809

However, the incorporation was carried out by mouldboard plough at the end of the rainy season preceding the millet crop, giving time for substantial mineralization of organic residues before planting and producing a physically improved rooting medium (see Chapter 11). For the ordinary farmer, unless he has ox-drawn equipment, there is no alternative to burning.

(c) *Crop residues*

Crop residues are frequently utilized off the field, e.g. groundnut haulms for livestock feed, cereal stems for fencing, but unsuitable or unwanted residues are burned on the field. This conserves nutrients (except N and S) where they are needed and leaves them available for the next crop.

3 ORGANIC MANURES

The traditional farmer appreciates the value of organic manures. Domestic wastes and animal droppings, as well as ashes, are usually collected and applied to the more frequently cropped "garden" land around the homestead. Netting (1968) described the importance of composted goat-pen manure in the agriculture of the Kofyar hill-farmers of Nigeria. Cattle-rearing, however, is nomadic or semi-nomadic and is little integrated with arable cropping, so that little, if any, farmyard manure is made and applied. The soil is manured when cattle are penned on crop-land by private arrangement between herdsman and cultivator, but on a regional scale the level of manuring achieved by this means seems likely to be very low.

Nevertheless, a great deal of experimental work has been done with farmyard manure. Yield responses have often been very large. Delbosc (1965), for example, reported that 5 or 10 t manure per ha applied to a two-year sorghum-groundnut rotation gave yield increases of up to 800 kg/ha groundnuts and 1700 kg/ha sorghum, an improvement of up to 200 per cent over control.

However, the precise nature of the benefit conferred on the soil by organic manures remains controversial. It is considered here under two broad headings: (a) physical and physico-chemical effects, and (b) provision of crop nutrients.

(a) *Physical and physico-chemical effects*

The addition of organic manure increases, at least temporarily, the soil organic matter content, initially that of the unhumified fraction, later and less certainly that of the humified fraction. The relationship between organic matter and soil physical properties has been discussed in Chapter 5, where it was concluded that the physical improvement that usually follows fallowing or incorporation of organic materials is probably due to the increase in the bulk of the unhumified fraction. The improvement is only temporary because of rapid decomposition, and improved soil structure persists only if fresh material is added annually. In addition, there was some evidence that an increase in humus content improved soil water-holding capacity to a small extent.

Two examples from the field may be cited. In dry areas of Senegal and Niger, the incorporation of 5—10 t/ha of crushed groundnut shells markedly increased subsequent yields of groundnuts and millet (Gillier, 1964; Lienart and Nabos, 1967), an effect attributable to better water utilization, either from improved infiltration or from increased water-holding capacity. And under higher rainfall in Ghana, yields were improved by surface mulching, due probably to better rainfall acceptance, lower soil temperatures, and a reduction in the capping and drying out of the soil (Nye, 1952b; Djokoto and Stephens, 1961a, b).

Farmyard manure is known to reduce the acidification of the soil that comes from continued cropping, particularly where ammonium fertilizers are often used (Djokoto and Stephens, 1961a; Ofori and Potakey, 1965; Dupont de Dinechin, 1967d; Pieri, 1971). This may be the result of the large base content (Ca, Mg, K and Na) of the manure or, as Bache (1965) suggested, of its incorporation of aluminium into insoluble complexes. Whatever the mechanism, the effect on crop yields where soils are already acid, can be large. Bache and Heathcote (1969) reported that in a long-term experiment at Samaru, increased cotton yields from treatments including manure coincided with higher soil pH, lower levels of soluble aluminium and manganese in the soil and lower levels of manganese in plant tissue (Table 44).

TABLE 44

Mean seed cotton yields from the DNPK experiment, Samaru, Nigeria, 1962-4 (Bache and Heathcote, 1969).

*Treatment**			*Cotton yield, kg/ha*	*Soil properties ***			*Plant Mn-content, ppm*
D	*N*	*P*		*pH*	*Al*	*Mn*	
0	2	2	432	4·00	0·47	0·26	245
1	1	1	798	4·29	0·10	0·11	149
2	0	0	935	4·64	0·03	0·07	95

*Fertilizer rates: D (manure), 0, 2·5 and 5·0 t/ha/annum
N (as amm. sulphate), 0, 12 and 24 kg N/ha/annum
P (as single s/p), 0, 5 and 10 kg P/ha/annum

** pH measured in a 1:1 0·01 M $CaCl_2$ suspension ; soluble Al and Mn values ($M \times 10^4$) measured in the filtrate.

Finally, any increases in soil organic matter from organic manuring raise the cation exchange capacity, and although this has no direct effect on crop growth it almost certainly makes the control of nutrient supplies to the crop easier.

(b) *Provision of plant nutrients*

Direct responses to farmyard manure have been recorded from rates as low as 2·5 t/ha (Greenwood, 1958), and Nye (1953) reported yields 50 per cent higher on fields that had received 5 t/ha within the previous three years. At such rates the effect is almost certainly chemical rather than physical and the consequence largely of the available phosphate content of the manure

(Hartley, 1937; Greenwood, 1948). Nye (1952b) reported a high negative superphosphate x manure interaction at phosphate deficient sites, although at low rates the effects of organic and inorganic sources may be additive (Nye, 1953).

As mentioned above, farmyard manure is also an important, if variable, source of cations (Table 45), and Bache and Heathcote (1969) reported increased levels of soil exchangeable cations from its continued use. Potassium is perhaps the most important cation supplied, and improved potassium nutrition following farmyard manure applications has been reported for cotton (Richard, 1967; Braud, 1971) and groundnuts (Prevot and Martin, 1964). In Ghana, sulphur responses were obtained on farmers' fields but not at experimental sites where manure had previously been applied, a result which suggests that the sulphur contribution of the manure may have been important (Stephens, 1960a); but sulphur deficiency was not corrected by farmyard manure in Upper Volta (Prevot and Martin, 1964).

On an equal nutrient content basis, organic manures are generally less effective short-term sources of plant nutrients than inorganic fertilizers (Hartley, 1937; Greenwood, 1948; Prevot and Martin, 1964), for elements held organically must be mineralized before they become available. This property of slow release may sometimes be an advantage. Nye (1952b) observed that, although superphosphate was needed for quick-growing crops, manure was usually a better source of phosphorus for long-season crops.

How much manure is required to maintain fertility under continuous cropping depends on the level of fertility desired. Amounts of about 5·0—7·5 t/ha in each three year rotation have been suggested (Dennison, 1959; 1961; Goldsworthy and Meredith, 1959), but these appear sufficient to maintain cereal and groundnut yields of only about 1000 kg/ha, equal perhaps to those obtained without fertilizer after a good bush fallow. On the other

TABLE 45

Some analyses of farmyard manure.

Element (% of dry matter)	*Origin*			
	Mali[1]	*Upper Volta*[2]		*Senegal*[3]
		pen manure	*farm manure*	
N	0·48—1.43	1·95	1·40	1·23
P	0·06—0·57	0·35	0·26	0·20
K	0·77—1·87	2·62	1·78	0·39
Ca	—	1·31	0·75	1·31
Mg	—	0·64	0·41	0·69

Data sources: 1 Pieri, 1971; 2 Prevot and Martin, 1964; 3 Delbosc, 1965.

TABLE 46

Maize yields (kg/ha) as affected by fertilizer and previous manuring on a twenty-year-old field at Samaru, Nigeria (Abdullahi, 1971).

*Fertilizer (kg/ha)**			*Previous long-term treatment, farmyard manure, t/ha/year.*			
N	*P*	*K*	0	2·5	7·5	12·5
0	0	0	33	584	2543	3145
134	28	56	1016	2316	3775	3821
268	56	112	2065	3311	4108	4247
** Soil organic C % :—			0·22	0·34	0·60	0·82

* Inorganic fertilizer applied in 1970 as calcium ammonium nitrate, single superphosphate and potassium chloride after 20 years manuring at four levels of application.
** from Jones, 1971.

hand, after 12·5 t/ha manure for 20 years, maize at Samaru yielded over 3000 kg/ha (Table 46), showing that a high rate of manuring can build up fertility.

As has been said, the ordinary farmer is not able to apply manure at such high rates. It is more realistic to recommend that he uses as much as is available and supplements with mineral fertilizers (Prevot and Martin, 1964). In any case, the nutrients in the manure are not always balanced with respect to the crop being grown. Nitrogen will often be insufficient for a good cereal crop, and existing soil deficiencies of other elements may be reflected in the chemical composition of the manure (Pieri, 1971).

(c) *Conclusions*

Of the organic manures, only farmyard manure has been widely tested, and its agricultural value is unquestioned. However, since manure is generally scarce, and only fundamental changes in the present pattern of agricultural activity can change this situation, it is of great practical importance to understand the nature of the benefits it confers, so that similar results may be achieved by other means.

Comparisons of manure with mineral fertilizers have usually shown the long-term advantage to lie with manure. Most probably, this advantage arises from the simultaneous contribution the manure makes to the physical, physico-chemical and nutritional aspects of fertility. In addition to supplying major nutrients, particularly N, P and K, and some micronutrients, farmyard manure may also, temporarily, effect an improvement in the physical state of the soil and, more permanently, augment the exchange complex and buffer the soil against acidification. Fortunately, there is good reason to think that this multiple action of manure can be satisfactorily simulated by management systems which combine the use of mineral fertilizers with judicious liming and the incorporation of all available residues.

4 INTERCROPPING

As was pointed out in Chapter 3, intercropping and sequential planting are widely practised, and study has shown that there are good economic reasons for doing so, but the effects of these practices on soil fertility have received little attention. However, some of these effects may be deduced by examination of the cropping sequence.

The commonest form of intercropping is probably semi-sequential planting, in which the second and subsequent crops are planted after the first, early-planted crop is established and frequently remain on the ground long after it has been harvested. Thus the soil carries plant cover for a longer period than would be the case with a sole crop. In addition, total plant population and total leaf area during the time when all the crops are growing together may be higher than those of a sole crop. One can therefore postulate that with mixed crops as compared with sole crops:

(i) the soil is better protected against the impact of rain, and consequently surface compaction and erosion are reduced;

(ii) transpiration is greater, so that downward leaching of nutrients is reduced;

(iii) exploitation of the soil nutrient supply at different depths is more efficient and extends over a longer period of time, so that total nutrient uptake is greater;

(iv) greater dry matter production leads to a greater production of crop residues, both above and below ground.

Support for the third of these assertions is provided by the results of an inter-row cropping experiment described by Andrews (1970). There were three crop treatments; two involved long-season sorghum grown in alternate rows with a short-season cereal followed by cowpeas, while in the third, sorghum was grown as a sole crop. Yields were much higher in the intercropped treatments (Table 47), and it may reasonably be assumed that these higher yields were accompanied by greater nutrient uptake.

TABLE 47

Grain yields from an inter-row cropping experiment (Andrews, 1970).

Treatment			*Yield, kg/ha* *Individual crops*	*Totals*
Intercrop:	(a)	Dwarf sorghum	1301	
	(b)	Millet,	2797	
		then cowpeas	210	4308
Intercrop:	(a)	Dwarf sorghum	1180	
	(b)	Maize,	3091	
		then cowpeas	149	4420
Sole crop:		Dwarf sorghum	2298	2298

Cereals may benefit from association with a legume. Steele (1972) at Kano reported that millet interplanted with cowpeas yielded more than sole crop millet grown at the same plant density. Andrews (1970) concluded that legumes are probably a necessary rotational component under most conditions, and certainly most indigenous crop rotations include a grain legume. There is scope here for further experimentation.

5 PASTORALISM

The savanna supports considerable numbers of cattle, sheep and goats, a large proportion of which are owned by nomadic Fulani. De Leeuw (1966) estimated the cattle population of northern Nigeria to be about eight millions.

Nomadic practices vary, and Allan (1965) referred to a report of Fulani herders in the grasslands of Guinea who practise "shifting pasturage", overgrazing one area for a few years before moving on. The destructive effect this has on the vegetation leads to an increased risk of erosion. But elsewhere movements of cattle are frequent, and over-grazing is probably serious only in the vicinity of Fulani camps, sedentary settlements and around watering points (de Leeuw, 1966). Well-used cattle trails may, however, initiate gully erosion in some areas.

In the past cultivators have welcomed the kraaling of nomadic cattle on their farmland, recognising the benefits of the dung. However, increases in human population and the introduction of cash crops have caused an encroachment of cultivation into the regular grazing grounds of the cattle-owners. This has led to friction between cultivators and pastoralists and to a decrease in the number of nomadic herds in the more densely populated areas, for example in the Sokoto and Kano provinces of northern Nigeria (de Leeuw, 1966). This is a problem which seems likely to spread as the number of people increases.

There is no information on the effects of the nomadic herds on soil fertility, but probably their main benefit to the farmer is that they convert into dung the bulky crop residues which are otherwise difficult to incorporate in the soil. Similarly, there is little information on the effects of soil fertility on the productivity of the animals, and although there is great scope for better quality forage, nomadism makes improvements difficult to introduce.

As well as the nomadic herds some cattle are kept by settled Fulani, and many farmers keep a few goats, sheep, and a donkey or mule for carrying goods. The dung from these animals is returned to the soil, as is that from animals kept in towns.

6 SOIL NUTRIENT BALANCE

The build-up of soil fertility under bush fallow has already been discussed, but since agriculture in the savanna is a cyclic process, cropping alternating with fallowing, the maintenance of fertility depends on the existence of a balance between nutrient accumulation under fallow and nutrient decline under cropping.

Under true subsistence farming there is considerable recycling of nutrients. Little, if any, of the harvest is exported from the neighbourhood of the farm; crop residues are burned on the field; and ash, household refuse and animal dung are returned to the field. By using manure and domestic residues in those areas immediately adjacent to the homestead, there is some transference of fertility away from the outlying areas, but in the long term the shifting of village sites has a levelling effect.

(a) *Losses*

The only significant pathways of nutrient loss from the system are (i) erosion of topsoil, (ii) loss in runoff water, (iii) leaching of cations and nitrate and sulphate out of the soil profile, (iv) the volatilization of nitrogen, sulphur and possibly phosphorus compounds through burning, and (v) removal in crops and stock sold off the farm.

(i) No figures are available for erosion losses from farmers' fields, but data from runoff plots at two sites, Sefa, Senegal, and Samaru, Nigeria, give a measure of losses under experimental conditions.

At Sefa, the mean soil loss from eight plots over ten years was 9·3 t/ha/annum, equivalent to one centimetre every 16—17 years (Roose, 1967).

At Samaru, erosion losses over four years averaged nearly 20 t/ha/annum from ridged land but were much less from land cultivated flat; for all cultivation treatments the average loss was 10 t/ha/annum (Kowal, 1970c). This amount of eroded soil contained about 6 kg/ha/annum of cations (Na, K, Ca and Mg) and 6 kg/ha/annum of nitrogen. Kowal, however, emphasized that his results represented "safe" conditions (in a soil conservation sense), the runoff being from graded bench terraces 183 m long with an average slope of only 0·3 per cent. Certainly many farms are on steeper sites, and losses on such farms may locally be much more severe.

(ii) Losses of nutrients in the runoff water were also measured in the Sefa and Samaru experiments. For all cultivation treatments the average annual losses over four years at Samaru were:

	Na	*K*	*Ca*	*Mg*	*N*
Nutrients in runoff water and suspended material (kg/ha)	1·6	6·3	4·9	2·1	7·4
Nutrients in eroded soil (kg/ha)	0·3	1·1	3·2	1·2	6·3

One may conclude that in these fertilized experimental plots, subject only to slight or moderate soil erosion, losses of nutrients in the runoff water exceed those in eroded soil. Comparison of soil analyses made before and after the four-year period showed that most of the nutrient losses could

TABLE 48

Leaching losses from the lysimeters at Bambey, Senegal in 1959 under 450 mm rain (Vidal and Fauche, 1962).

Soil cover	*Drainage, M^3/ha*	*Nutrients in drainage water, kg/ha*				
		N	*P*	*K*	*Ca*	*Mg*
Millet	952±22	1·6±0·2	0·06±0·01	3·2±0·2	17·9±0·6	2·6±0·2
Natural fallow						
—incorporated	967±86	6·3±1·1	0·07±0·01	5·7±0·5	22·0±1·2	12·1±2·4
—burned	913±76	6·6±0·7	0·04±0·01	5·1±0·5	20·9±1·4	6·1±1·2
Bare fallow	1393±26	45·6±0·8	0·13±0·02	11·5±0·6	55·4±3·2	23·3±2·0

be accounted for in terms of runoff and erosion; that is, leaching losses were low.

(iii) Throughout the savanna through-drainage occurs in years of average rainfall. In a laboratory experiment on Senegal soils, Moureaux and Fauck (1967) showed that leaching removed exchangeable bases and increased soil acidity, the effects being most marked on sandy soils; and in East Africa, Scott (1962) traced a relationship between base saturation of the soil and the annual rainfall.

Nye and Greenland (1960) and Vine (1968) pointed out that losses of cations from soils are necessarily dependent on losses of anions, and that under the conditions of the humid tropics the only important anion is nitrate. Hence, significant leaching of cations occurs only when the production of nitrate in the soil exceeds its uptake by plants and micro-organisms. Particularly in the northern savanna where fallows are predominantly of grass, soil nitrate levels are low during the fallow and for a time after it has been cleared, and losses of cations during this period are likely to be small. But in the later stages of the cropping cycle, when soil C:N ratios are lower, the excess of nitrate production over uptake may be sufficient to promote appreciable leaching of cations. The same is likely to happen when nitrogen fertilizers are used.

The density of plant cover is important. Vidal and Fauche (1962) showed that on a sandy soil in Senegal leaching losses of mineral elements were much greater under bare fallow than under bush fallow or crop cover (Table 48). These losses, it should be noted, were calculated from analyses of drainage water from lysimeters 40 cm deep and were probably greater than occur under field conditions, where at least some crop roots grow to greater depths. Practices such as mixed cropping and semi-sequential planting are beneficial, for by keeping the land well covered they serve to minimise leaching losses.

(iv) The burning of bush, crop residues and firewood removes nitrogen, sulphur and possibly other nutrients from the soil-plant system. As noted above, Vidal and Fauche (1961; 1962) reported that burning a fallow entailed a nitrogen loss of up to 20 kg/ha; but others have found an increase in total and available nitrogen after burning, which in some environments is partly the result of enhanced fixation by legumes (Daubenmire, 1968). There is, however, no similar mechanism for compensating for the loss of sulphur by burning.

(v) Losses of nutrients through cropping become serious when crops are sold off the farm, and, because of low levels in the soil, that of phosphate is the most serious. With the yields commonly attained by farmers actual removal of phosphate per hectare is small (Table 49), but Watson (1964) estimated that for northern Nigeria about 25,000 tonnes of single superphosphate were needed annually to replace the phosphate removed in the groundnut crop sold for export. The corresponding figure for cotton seed

TABLE 49

Amounts of phosphorus in crops sold off the farm*

Crop	*Yield (kg/ha)*	*Loss of P (kg/ha)*
Groundnuts (kernels)	500	2
Cotton (seed)	250	1
Cereals (grain)	1000	3

* from Watson, 1964, and Table 57.

TABLE 50

Nutrients in rainwater.

	Rainfall nutrient addition, kg/ha/annum				
Site	*P*	*K*	*Ca*	*Mg*	*Na*
1. Samaru, Nigeria	2·6	36·7	1·1	2·9	60·3
2. Zaire (mean of three sites)	—	2·2	4·2	1·1	—
3. Gambia a	0·3	5·9	4·4	—	9·5
b	0·2	4·3	2·7	—	8·9
c	0·3	2·8	1·7	—	5·8
4. Katherine, N.T., Australia	—	0·3	—	0·2	1·1

Data sources: 1. Jones, 1960; 2. Meyer and Dupriez, 1959; 3. Thornton, 1965, sites a, b and c were 14, 130 and 219 km. from sea, respectively; 4. Wetselaar and Hutton, 1963.

was 2,000 tonnes. Nutrients, including phosphate, are also removed by stock, but under the usual extensive grazing system the removal per hectare must be very small.

(b) *Gains*

Ways in which nutrients are added to the agricultural system include (i) microbiological fixation of nitrogen (see Chapter 6), (ii) rainfall, (iii) deposition in dust, (iv) weathering of profile minerals and parent material, and (v) fertilizers.

There have been only a few chemical analyses of rainwater in West Africa. Nitrogen data were discussed in Chapter 6 and sulphur data in Chapter 8. For other nutrients, reports from the savanna and other areas of similar climate are summarized in Table 50. Apart from the high values of Jones (1960), which might have been affected by contamination, there is little to suggest that rain makes more than a small contribution. Thornton (1965) pointed out that potassium and calcium concentrations were higher at the beginning of the rains and suggested that these elements are either washed from, or fall in the form of, dust present in the atmosphere at this time. High concentrations of sodium in the early rains in Gambia were probably related to fine particles of salt water carried in from the sea. For similar climatic conditions in Australia, Wetselaar and Hutton (1963) argued that most of the ions dissolved in the rainwater were part of the terrestrial cycle and could not be regarded as a true accession to the soil, and Meyer and Dupriez (1959) advanced a similar view about rainfall nitrogen in Zaire; but the results of Thornton (1965) demonstrate a maritime influence in the western savanna.

Analysis of harmattan dust in Nigeria by Doyne *et al.* (1938) showed it to be formed largely of diatoms and fine particles of quartz. A large part of the material was in the clay fraction, and this consisted mostly of silica, alumina, and ferric oxides. Significant amounts of calcium, magnesium, potassium and sodium were also present, but no phosphate was detected. However, the recent report of Bromfield (1974a) that dust deposition along a 500 km transect of the Nigerian savanna amounted to only 52—226 kg/ha/annum throws doubt on previous assumptions of a fertilizing effect as far south as the coastal forests (see Hamilton, 1966).

The contribution to maintenance of fertility by the weathering of primary minerals in the soil profile is not easy to assess. Nye (1963) stated that "most tropical soils are either too deeply weathered for the roots of crops to be in contact with primary minerals, or are derived from sediments in which the minerals that have survived are necessarily very stable". The roots of bush fallow vegetation, however, normally go deeper than those of annual crops; and while in some southern areas soils may be so deeply weathered that even the roots of trees make little effective contact with weathering parent material, much of the savanna is covered by relatively shallow soil (Maignien, 1961; Fauck, 1963; d'Hoore, 1964; Klinkenberg and Higgins, 1970), in which the zone of weathering is more accessible. How much this zone can contribute to the fertility of the topsoil is not known in quantitative terms but is certainly dependent on the quality of the parent material. Under subsistence farming the long-term maintenance of soil fertility requires that losses from the topsoil by erosion and leaching are made good by contributions from the subsoil.

(c) *Fertilizers and nutrient balance*

Present consumption of fertilizers is low but increasing. The pattern now and for the foreseeable future is for much of it to be used at fairly low rates in those areas where the traditional system has already been adapted to include a cash crop (e.g. phosphate for groundnuts). Such use often leads to an extension of the cropping cycle at the expense of the fallow period, and by providing only those nutrients that are most limiting (P and sometimes N or K), it inevitably imposes a strain on supplies of other nutrients, particularly K (where not supplied), Mg, S and the micronutrients. A second point concerns nitrogen fertilizers. Even when ammonium forms and urea are used at low rates, some increase in the nitrate content of the water leaching through the profile, and therefore in the loss of cations, is probably unavoidable.

Hence, under intensive cultivation, fertilizers which supply only one or two nutrients may harm the nutrient balance of the soil. There is as yet little evidence of nutrient imbalance, which would manifest itself by an increasing number of nutrient deficiencies; but it must be considered as a strong long-term possibility because of the very low cation exchange capacities of the soils. If the problem does arise, balanced fertilizers containing several nutrients will be needed, a point referred to again at the end of Chapter 10.

PART III—FERTILITY REQUIREMENTS FOR HIGHER CROP YIELDS

The increasing demand for agricultural products will almost certainly be met by increases in production by small farmers. To some extent these increases can be achieved by bringing more land into cultivation. Uncultivated land exists, but it is not always available to the farmer or easily accessible, and even where it is, it may be poorer than that used at present. Increased production can probably be achieved more easily and more cheaply by increasing the yield per hectare through modifications to the traditional form of agriculture.

Undoubtedly, yields can be raised by technical improvements in farming methods. These improvements are well known and may be summarized as:

(i) planting at the correct time and optimal density;

(ii) better control of weeds, pests and diseases;

(iii) better soil cultivation and conservation;

(iv) better crop varieties, more responsive to higher levels of fertility;

(v) use of fertilizers;

(vi) irrigation (at suitable sites).

But certain qualifications are needed. First, improvements in yield potential must not increase the risk of crop failure in bad years; secondly, there must be an assured market for the extra produce, which implies the availability of adequate storage and transport facilities; and thirdly, the farmer must get sufficient return from his extra labour and cash outlay. Unless these requirements are met, no technical innovation is likely to make much impact on production.

Further, the interdependence of technical innovations must be stressed. None is likely to be very effective in isolation. There is no point in improving and maintaining soil fertility if, for instance, the lack of responsive crop varieties, effective disease control or adequate weeding prevents it from being reflected in improved yields. In most situations it is necessary to introduce innovations together as a "package deal" (e.g. fertilizer, improved seed, and advice on cultural techniques), as was demonstrated by Allan (1968, 1971) in Kenya.

The present study is concerned with soil fertility and its maintenance, and these broader agronomic issues are beyond the scope of our discussions. Nevertheless, their existence should be kept in mind as, in the next two chapters, we examine, first, the evidence from numerous fertilizer trials and, secondly, the data from a more limited number of experiments on soil cultivations and the conservation of soil and water, and try to relate them to the needs of the farmer.

Chapter 10

NUTRIENT REQUIREMENTS: THE EVIDENCE FROM FERTILIZER EXPERIMENTS

The general pattern of soil fertility outlined in earlier chapters was based on chemical and pedological data. However, the ultimate test of fertility is the growth of the crop itself. The present chapter therefore considers fertility by reference to the results of numerous field trials. It reviews nutrient deficiencies and the pattern of their occurrence, and examines the prospects for increasing and maintaining fertility through the use of inorganic fertilizers.

A soil may be considered fertile if it can support optimal growth of the crop it carries. However, crops differ and some, particularly new, higher yielding varieties, are more demanding than others. Moreover, the fertility of a soil changes with time: a soil that was fertile the first season after a long bush fallow is unlikely to remain so under prolonged unfertilized cropping, although the rate of decline in fertility and the order of appearance of individual nutrient deficiencies will vary with soil type and the nature of the crop rotation grown.

We therefore look first at the initial nutrient deficiencies, as shown by response to fertilizer in the first year or two of cropping, in relation to soil, climatic and agronomic factors. Much of the data comes from one-year or short-term experiments, set up primarily to establish economic fertilizer recommendations for farmers, and usually carried out with unimproved crop varieties and using traditional cultural practices. We have termed this low intensity farming.

Secondly, we consider the evolution of soil fertility under continuous cropping and the deficiencies that appear when higher yielding crops are grown in more intensive farming systems. The data here come from a much smaller number of cropping and rotation experiments designed to study long-term effects.

1 NUTRIENT DEFICIENCIES AND FERTILIZER RESPONSES AT LOW FARMING INTENSITY

(a) *Initial fertilizer responses*

(i) *Phosphate*

By far the most frequent response has been to phosphate. With groundnuts phosphate responses occur throughout the region and may often be very large. However, exceptions have been reported (a) on an alluvial soil in Gambia (Evelyn and Thornton, 1964), (b) in the Thies rock-phosphate zone (Bockelee-Morvan, 1964; 1965a), (c) in the Patar potassium-deficient zone in Senegal, and (d) at a severely sulphur-deficient site in Nigeria (Greenwood, 1954). The degree of response may be limited in some areas by less acute sulphur-deficiency (Ollagnier and Prevot, 1958a; Stephens, 1960a; Goldsworthy and Heathcote, 1963; Bockelee-Morvan, 1966) or by low rainfall (Prevot and Ollagnier, 1959; Lienart and Nabos, 1967). It seems probable that similar remarks apply to soyabeans (Goldsworthy and Heathcote, 1964; Watson, 1964) and cowpeas (Nicou and Poulain, 1967).

Rainfed cereals also respond to phosphate at most sites, although response may be small or irregular in more humid areas, where nitrogen is the major limiting nutrient (Stephens, 1960a; Goldsworthy, 1967a; Bromfield, 1967). All cereals, and in particular maize, have a high nitrogen requirement which is not fully met by the annual mineralization of soil organic nitrogen. Hence, applications of phosphate to cereals are usually more effective in the presence of applied nitrogen (Nye, 1953; 1954; Charreau and Poulain, 1963; Bouchet, 1963; Meredith, 1965; Goldsworthy, 1967a; 1967b; Pieri, 1971; IRAT/Haute Volta, 1972). Positive N x P interactions have been reported from Mali (Bono, 1970), Ghana (Nye, 1952a) and northern Dahomey (Werts and Bouyer, 1967).

Responses of rice to phosphate tend to be smaller than those of other cereals under both rainfed (Stephens, 1960a; Bourke, 1965) and flooded conditions (Dumont, 1966; Bredero, 1966). In the latter circumstances this may be due either to the intrinsically high phosphate status of certain valley-bottom soils (Pieri, 1967a; Dumont and Dupont de Dinechin, 1967) or to low fertilizer efficiency because of strong phosphate adsorption (Couey *et al.*, 1967). Irrigated wheat, grown at four sites in northern Nigeria, showed no significant response to phosphate (Andrews, 1968).

Phosphate has been reported as the first limiting nutrient for cotton in northern Dahomey (IRCT, 1965a), Mali (IRCT, 1965b; Pieri, 1967a), Upper Volta (Jenny, 1965) and Nigeria (Watson, 1964), but in Chad (Megie *et al.*, 1970) and Ivory Coast (Bouchy, 1970) phosphate, though necessary, is secondary in importance to nitrogen and sulphur in the early years after bush clearance.

(ii) *Nitrogen*

Although the most important responses occur in cotton and cereal crops, nitrogen may also increase groundnut yields, particularly the haulm (Heathcote, 1970) and particularly in very dry areas (Bockelee-Morvan, 1965a; 1965b; 1966), an effect which is independent of that of sulphur where ammonium sulphate is the nitrogen source.

Most cereals show some response to nitrogen, but this element is the first limiting nutrient only in some of the more humid parts of the region (Bromfield, 1967), where long grass fallows have just been cleared (Nye, 1951; Stephens, 1960c; Djokoto and Stephens, 1961a) or on irrigated rice (Bourke, 1965; Jenny, 1965; Dumont, 1966). Elsewhere, phosphate deficiency must be remedied before applied nitrogen can give substantial yield increases. Even then some long-season varieties show little response. Stockinger (1970), comparing maize and long-season sorghum, concluded that the greater response of maize might be due to its shorter growth period and hence greater intensity of demand. The introduction of more high-yielding short-season cereal varieties will intensify rates of uptake and, probably, response (Blondel, 1971b).

Cotton commonly responds to nitrogen, but the effect may be confined to dry-matter production where planting is late or insect-control lacking (Watson, 1964; Palmer and Goldsworthy, 1971), and it may be limited by shortage of phosphate (Jenny, 1965) or boron (Heathcote and Smithson, 1974). The response is most marked immediately after fallow clearing, and in Ivory Coast and Chad nitrogen is the main limiting nutrient for the first few years (IRCT, 1965c; Bouchy, 1970; Megie *et al.*, 1970).

(iii) *Potassium*

Potassium has been widely reported as affording little or no benefit in non-intensive savanna agriculture (Nye, 1952b; Dennison, 1959; Jenny 1965; Werts and Bouyer, 1965; 1967; Amagou, 1965; Dupont de Dinechin, 1967c; Pieri, 1971); and there have been suggestions of depressive effects on sorghum in northern Cameroun (Vaille, 1970) and on groundnuts in Ghana (Nye, 1952b). However, in the west of the region legume crops have responded positively to small applications of potassium. In Gambia, haulm responses of groundnuts were recorded at all four sites tested and a positive nut response at one of them (Evelyn and Thornton, 1964). In Senegal, although deficiency was described as rare by Ollagnier and Prevot (1958a), small dressings are now thought to be useful in the centre and south of the groundnut belt (Bockelee-Morvan, 1964; 1966; Poulain and Arrivets, 1972) and are essential in the small potassium-deficient zone around Patar and in the rock-phosphate zone at Thies (Bockelee-Morvan, 1964; Gillier and Gautreau, 1971). Cowpeas have a higher potassium requirement than groundnuts, and fertilizer potassium is regarded as essential for this crop in some parts of Senegal (Nicou and Poulain, 1967).

From the derived savanna of Nigeria Bromfield (1967) reported a small and irregular response of maize to potassium, more noticeable in droughty years; Bouchet (1963), in Mali, stated that potassium was useful for millet and sorghum on soils of schistose origin; and Poulain and Arrivets (1972) reported sorghum responses on Ferruginous Soils and a Vertisol in Upper Volta; but no other reports of cereal responses to potassium have been found. On cotton, responses are very small or non-existent on recently cleared soils, but may thereafter appear quite quickly (Bouchy, 1970; Megie *et al.*, 1970).

(iv) *Sulphur*

Deficiency of sulphur, particularly in legumes, has long been recognized (Greenwood, 1951; 1954), but through the widespread use of single superphosphate and ammonium sulphate this element has usually been oversupplied at most experimental sites and has therefore been ignored. Now that the trend is towards the use of higher analysis sources of phosphorus and nitrogen containing little or no sulphur, the sulphur status of savanna soils needs reassessment.

The existence of a sulphur requirement has usually been shown by a negative interaction between single superphosphate and ammonium sulphate, or by direct yield response to gypsum or ammonium sulphate. Although with ammonium sulphate there is risk of confusion with a nitrogen response, and with gypsum a calcium response (Russell, 1968a), there is little doubt that the deficiency is a real one. There have been reports from many areas of groundnut responses to sulphur (Nye, 1952a; Greenwood, 1954; Ollagnier and Prevot, 1958a; Stephens, 1960a; 1960b; Bolle-Jones, 1964; Bockelee-Morvan, 1965b; 1966), and one of soyabeans (Goldsworthy and Heathcote, 1964).

Cereal crops are less demanding, but Charreau and Poulain (1963), in Senegal, attributed the greater response of millet to ammonium sulphate, as compared with urea and calcium nitrate, to its sulphur content. With cotton, deficiencies have been reported from Chad, Central African Republic and Ivory Coast, being most severe just after bush clearing (Roux and Gutknecht, 1956; Bouchy, 1970; Megie *et al.*, 1970).

(v) *Other elements*

Responses to elements other than N, P, K and S have been rare. Deficiency of calcium was blamed for low shelling percentages of groundnuts at two sites in Nigeria (Meredith, 1965) and, as the result of potassium fertilization, for low yields in Ghana (Nye, 1952b), but has not been recorded in Senegal (Ollagnier and Prevot, 1958a; Bockelee-Morvan, 1965b). Pot tests indicated magnesium deficiency in one Niger soil (Lienart and Nabos, 1967) and deficiency symptoms have been observed on maize at Grimari, Central African Republic (Morel and Quantin, 1972). Field trials showed no response to applied magnesium in Ghana (Stephens, 1959).

Small responses by groundnuts to molybdenum have been reported from Senegal (Ollagnier and Prevot, 1958a; Fourrier, 1965; Bockelee-Morvan, 1965b; Gillier, 1966), Ghana (Stephens, 1959) and Nigeria (Heathcote, 1972a); and there was one response to copper (Bockelee-Morvan, 1965b).

There was no response to applied Cu, B, Fe, Mn and Zn either on food crops in Ghana (Stephens, 1959) or groundnuts in Senegal (Fourrier, 1965). Response to B has been found in cotton on farmers' land in Nigeria (Heathcote and Smithson, 1974), and irrigated rice has responded substantially to applied Zn on Vertisols near Lake Chad (unpublished observations), suggesting that shortage of Zn may be severe in the area. Nevertheless, the present evidence is that although there are responses to the application of secondary and micro-nutrients to some crops in certain areas, they are very much less common at low farming intensity than responses to major nutrients.

(vi) *Liming*

Nye (1952b) reported a small but fairly consistent effect of liming on groundnuts in Ghana; Blondel (1970) recommended liming for certain dune-summit soils in Senegal, in which chlorosis in groundnuts appeared to be due to the inhibition of rhizobia by low pH; and Beye (1972) found that rice responded to heavy dressings of lime on acid sulphate soils in the Senegal river delta; but elsewhere, unless the soil has become acid from the use of ammonium fertilizers, liming has had little or no effect (Nye, 1952b; Dennison, 1959; Bockelee-Morvan, 1965a; Poulain, 1967; IRAT/Haute Volta, 1972). As Russell (1968a) has pointed out, this result contrasts with temperate experience. He suggested that lime may be of importance only on soils having a relatively high content of exchangeable or active aluminium. Many well-drained tropical soils may have low pH and yet also a low content of exchangeable aluminium because of their low exchange capacities. In these soils it may be necessary for the pH to fall to about 4·5 before mobile aluminium becomes important.

(b) *Fertilizer responses in relation to soil and climatic factors*

Interpretation of the fertilizer responses outlined above in terms of more fundamental environmental factors are severely hampered by the frequent lack of ancillary information on climate, soil type, parent material and the previous utilization and manuring history of the site.

However, it seems clear that phosphate is usually the main limiting nutrient in Ferruginous Soils and Ferrisols (Amagou, 1965), but is less likely to be so in Eutrophic Brown Soils, Vertisols and alluvial soils (Evelyn and Thornton, 1964; Dumont and Dupont de Dinechin, 1967); while in some Hydromorphic Soils, in spite of severe deficiency, response to phosphate may be restricted by high adsorption capacity (Jacquinot, 1965; Dumont, 1966; Couey *et al.*, 1965; 1967).

The degree of nitrogen deficiency is related more to organic matter levels and to the C:N ratio, and hence, indirectly, to the rainfall pattern and the vegetation type, than to the type of soil and parent material. However, response to added nitrogen, in particular the efficiency of its utilization, has been shown to vary with soil type (IRAT/Haute Volta, 1972) largely through differences in efficiency of water utilization.

Potassium deficiency is rare on newly cleared land, but may appear rapidly on some slightly Ferrallitic Soils, like those of the "Terres de Barre" in Dahomey and the Logone region of Chad (Werts and Bouyer, 1967; Braud, 1971). Elsewhere its appearance usually follows several years of fairly intense cropping but is likely to be earliest in soils overlying coarse sedimentary parent materials.

Crop growth is very sensitive to differences in soil moisture regime, and in areas of marginal rainfall poor infiltration and low moisture retention capacity may severely restrict fertilizer responses (Charreau and Poulain, 1963; 1964; Gautreau, 1965). A phosphate response dependent on rainfall has been observed in both groundnuts (Prevot and Ollagnier, 1959; Goldsworthy and Heathcote, 1963; Lienart and Nabos, 1967) and cowpeas (Nicou and Poulain, 1967), presumably as a result of water limitation. Where rock phosphate is used the effect of low rainfall is to limit phosphate dissolution and hence its availability. This source is not recommended where mean annual rainfall is less than 700—800 mm (Ollagnier and Prevot, 1958b; Bockelee-Morvan, 1965a; Poulain and Mara, 1965; Dupont de Dinechin, 1967b; Pieri, 1971).

In a few experiments fertilizer response has been enhanced by drought, presumably due to the fertilizer increasing root growth. Responses to phosphate and potassium were generally greater in droughty years in southern Nigeria (Bromfield, 1967), and further north at Samaru boron deficiency is most noticeable on soils that dry out quickly (Heathcote and Smithson, 1974). The opposite effect is more usual with nitrogen, which stimulates leaf growth and thereby increases the susceptibility of the crop to drought.

The tendency for yields and response to fertilizer to decline with lateness of planting appears to be general for most crops (Watson, 1964; Stockinger, 1970; Ganry, 1972; Haddad and Seguy, 1972). This decline can often be traced to the greater incidence of moisture stress or of pests and diseases on late-planted crops, but in some cases changes in soil physical properties and nutrient availability may also be involved.

The existence of both zonal and local variations in the patterns of fertilizer response and their interaction with environmental and human factors may be illustrated by two examples:

(i) During the period 1957-64 a large number of one-year trials were carried out on farmers' land in northern Nigeria to determine economic fertilizer rates for groundnuts, sorghum and maize (Goldsworthy and Heathcote, 1963; Goldsworthy, 1967a; 1967b), and from the results it was possible to divide the main areas of production into a small number of response zones, according to crop response to nitrogen and phosphate.

With groundnuts the response zones comprised four approximately parallel belts running east-west across the country. Control yields and response to superphosphate were least in the most northerly belt, probably due to a moisture limitation. In the others, response increased northwards, being greatest where the soil content of total and inorganic phosphate was least (Table 51), though the poor response in the south may have been due in part to rosette disease.

Different response zones were found for sorghum and maize. Again response to applied phosphate increased from south to north, but phosphate deficiency appeared to be most severe in an area of Ferruginous Soils developed on drift, or loess, centred on Zaria, and responses here were greater than those on other Ferruginous Soils over basement complex rocks. These results suggest that phosphate is less deficient in Ferrallitic than in Ferruginous soils, but the evidence is confused by the confounding between rainfall, and therefore soil organic matter, and soil type. Nitrogen responses also increased northwards, following the decline of organic matter and total soil nitrogen.

These findings may be compared with those of Nye (1953; 1954) in Ghana, who showed that, within a very broad belt of Ferruginous Soils, crop response to phosphate increased northwards from the Voltaian sandstone area in the south to the granite area in the extreme north. The effect appeared to be related to the shorter fallows prevailing in the north, but there was the possibility that the northern soils were inherently poorer in phosphate.

(ii) Differences in natural fertility and hence in fertilizer response may also show a much more local variation. Dumont and Dupont de Dinechin (1967) set up four NPK field experiments at different points along a soil catena at Tougou, Upper Volta. At the top of the slope, shallow soils over ironpan showed a very low natural fertility, and nitrogen, phosphate and water were limiting. Middle slope soils, which were deeper and more frequently cultivated, were very deficient in phosphate and potassium, but water

TABLE 51

Groundnut response zones in Nigeria (Goldsworthy and Heathcote, 1963).

	Response zones			
	1	2	3	4
Mean latitude	12° 30′	11° 30′	10° 20′	8° 30′
Rainy season, days	100	121	145	183
Rainfall, mm	400—800	700—1000	700—1400	1000—1400
Mean soil P, ppm				
— total	49	77	173	149
— inorganic	28	46	125	95
— organic	21	31	48	54
G'nut yields, kg/ha				
— control plots	416	933	987	896
— response, P_1*	50	136	90	62
— response, P_2*	41	187	129	67

*P_1=63 kg/ha single superphosphate; P_2=125 kg/ha single superphosphate.

was less limiting and fertilizer responses were high. In soils of the lower slopes and valley bottom fertility was higher, and the main problem seemed to be soil physical conditions. The authors attributed a general increase in phosphate availability down the slope in part to enrichment by alluvium and colluvium, and in part to less frequent cropping of the heavier valley soils.

(c) *Limitation of fertilizer response by agronomic factors*

In the groundnut trials of Goldsworthy and Heathcote, cited above, the mean crop response was only 1—2 kg groundnuts per kg single superphosphate. Such a result is not unusual. Low response, despite known deficiency, is a common feature of fertilizer trials conducted on farmers' fields. In some cases this may be due to deficiency of some other nutrient, but more often it is the result of poor husbandry. Work in Kenya (Allan, 1971) has shown that factors most affecting farmers' yields are planting date, weeding efficiency, crop variety and, only fourthly, crop nutrition. There is much circumstantial evidence that the situation is similar in the savanna, and other factors such as cultivation depth (Poulain and Tourte, 1970; IRAT/Haute Volta, 1972) and insect control (de Geus, 1967) are important at some sites.

With groundnuts in Upper Volta (Galland, 1963), yields were increased about three-fold from a control yield of 500—600 kg/ha by a combination of five treatments (i) use of 2—5 t/ha of farmyard manure and 75 kg/ha of single superphosphate, (ii) growth of a high-yielding variety resistant to rosette disease, (iii) use of seed dressed with fungicide and insecticide, (iv) high planting density (80,000 on ridges, or 110,000 seeds/ha on flat), and (v) good weed control. Similarly, in northern Nigeria (Harkness, 1970), groundnut yields were increased from 680 kg/ha under traditional cultivation to about 1300 kg/ha by using the recommended crop variety, dressed seed, planting at the correct time, using insecticide spray, and using 93 kg/ha of single superphosphate.

Work of this type gives no indication of the contribution of each agronomic improvement to the final yield, and more analyses of the kind carried out by Allan (1971) are needed; but, altogether, the effect can be very large. It is misleading therefore to judge the future of fertilizer use from responses where husbandry is poor.

One further example will be given. Responses of cotton to fertilizer have often been uneconomic in Nigeria. However, when Palmer and Goldsworthy (1971) compared the response of farmers' crops to nitrogen and phosphate in the presence and absence of insect control, they found that with no control of insects there was a fertilizer response only in the eastern zone of the country, but with control there were responses in all three zones (Figure 8). In these experiments the mean response to 125 kg/ha of ammonium sulphate was 33 kg/ha of seed cotton in the absence of insecticide spray and 106 kg/ha in its presence. There was a similar effect on the response to single superphosphate.

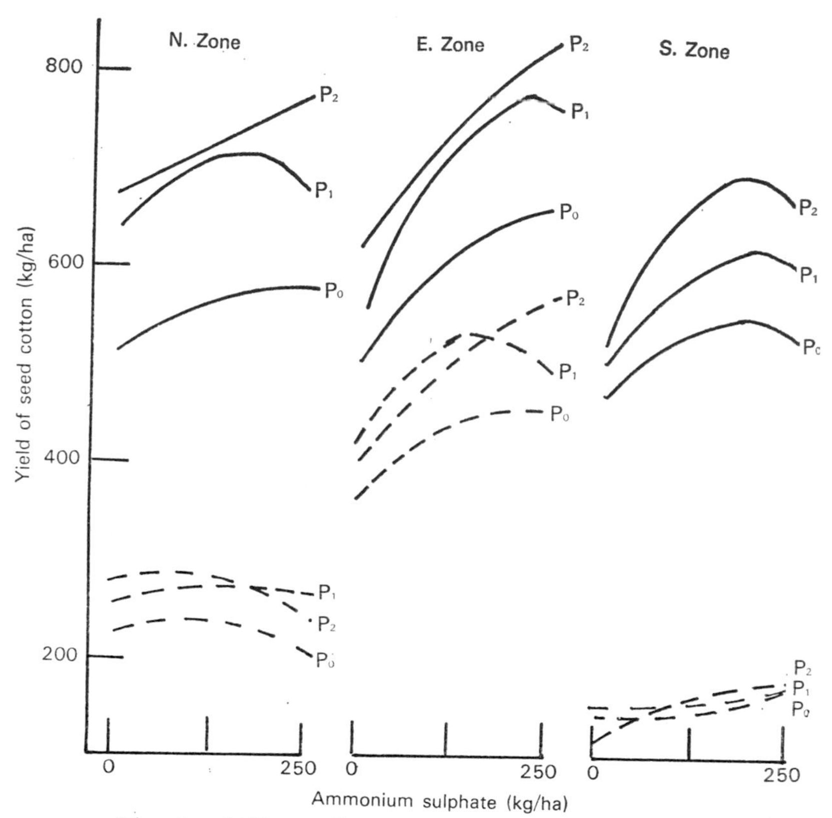

Fig. 8. Effects of ammonium sulphate at three levels of superphosphate on yields of seed cotton in Nigeria (Palmer and Goldsworthy, 1971).

——— sprayed with insecticide
- - - - - - unsprayed

2 NUTRIENT DEFICIENCIES AND FERTILIZER RESPONSES AT HIGH FARMING INTENSITY

Higher yields demand higher fertility, and this almost inevitably means a greater use of fertilizers. Various farming systems can be devised which give higher yields, and they need not necessarily exclude the traditional bush fallow; but higher yields do imply a more sophisticated form of farming, usually with a greater capital investment per hectare, in which the reversion of land to its natural vegetation would seem to have no place. The two aspects of high farming intensity, high annual productivity and continuous cultivation are therefore related.

However, we look first at the effect on fertility of prolonged cropping without fertilizer.

(a) *Effects of continuous cropping on soil fertility*

A decline in yields under continuous cultivation has frequently been reported and cited as the main reason for the adoption of shifting cultivation and bush fallowing. Studies of soil cultivated for a number of years have always shown large decreases, with time, in all the important parameters of soil fertility, notably organic matter (Djokoto and Stephens, 1961b; Charreau and Fauck, 1970; Jones, 1971), cation exchange capacity and exchangeable cations (Bouyer, 1959; Prevot and Ollagnier, 1959; Morel and Quantin, 1972). The rate of loss of cations is usually even greater than the decline in exchange capacity, so that base saturation and therefore pH also decline. Charreau and Fauck (1970) reported the loss of 45—50 per cent of the exchangeable bases and a pH drop from 6·4 to between 4·6 and 5·1 during 15 years cultivation at Sefa. These changes have been discussed in earlier Chapters.

Most observations of the effects of continuous cropping have come from experimental sites, but Siband (1972) described differences in soil characteristics and crop yields on farmers' fields of different ages at two sites in Senegal (Table 52). A one-year trial on these fields with rice as test crop was handicapped by late drought, particularly at Diankancounda Ogueil, but even the incomplete results show poorest growth on fields which had been longest in cultivation, despite large and uniform applications of mineral fertilizers. Notable features were the greatly prolonged vegetative period and the low rooting density on the oldest fields.

These results may be compared with those of a maize trial at Samaru, Nigeria (Abdullahi, 1971). Plots, previously cropped for 20 years with cotton, sorghum and groundnuts without mineral fertilizers but with four rates of farmyard manure, were split and sown to maize at three levels of NPK fertilizer. Yields were greatly increased both by the fertilizers and by the cumulative effects of twenty annual applications of farmyard manure, but the highest rate of fertilizer produced little more than 2000 kg grain per ha on unmanured plots, a very poor response by local standards (Table 46).

TABLE 52

Effect of length of cropping on soil properties and crop yields at two sites in Senegal (Siband, 1972).

Site:	*Sare Bidji*			*Diankancounda Ogueil*		
Soil type:	*Slightly Ferrallitic*			*Leached Ferruginous*		
Years since clearing:	5	16	80	2	15	50
Soil properties (0—10 cm):—						
Clay, %	9·1	—	6·0	9·3	—	6·5
Organic matter, %	2·4	1·5	1·3	1·5	1·3	1·1
C.E.C., me/100 g	4·7	3·5	2·6	3·5	3·2	2·3
Exch. Ca, me/100 g	2·9	1·7	1·1	2·1	1·5	0·7
Exch. K, me/100 g	0·09	0·07	0·08	0·10	0·08	0·04
Exch. Mg, me/100 g	1·0	0·9	0·5	0·8	0·6	0·4
pH (water)	6·5	6·4	6·0	6·6	6·2	5·6
*Field experiments, rice yields, kg/ha:—						
Grain**	2463	1404	0	0	0	0
Stover	2950	2990	880	2750	2640	1500
% K_2O in plants at flowering	2·24	1·97	1·60	—	—	—

*All fields received PKS fertilizer and 100 kg N/ha in two split dressings.
**No grain yield on oldest field at Sare Bidji. At Diakancounda Ogueil drought prevented grain formation in all three fields.

Both these experiments show a low efficiency for mineral fertilizers on land degraded by long cultivation with little or no fertilization. Unfortunately, neither was continued beyond the first season, so we cannot know whether further applications of fertilizers would eventually have restored a high level of fertility. With only a single fertilizer application in one year there is a strong possibility that the applied nutrients, and phosphate in particular, had a low availability because of poor physical distribution in the soil.

Very little is known about which factors most limit crop production on seriously degraded land, but these results suggest that deficiencies of major nutrients may not be the only ones involved. Soil physical properties may be of equal importance. Prolonged cropping has been shown to increase soil bulk density in Senegal (Charreau and Fauck, 1970), and high soil density could explain the poor rooting in the oldest fields in Siband's experiment and thus account for poor utilization of applied nutrients. Elsewhere in Senegal the response of sorghum to applied nitrogen was much increased by deep ploughing (Poulain and Tourte, 1970).

(b) *Changes in fertilizer response under continuous cropping*

As has been shown, fertilizer responses in the first few years after clearing land from fallow are almost entirely restricted to phosphate, nitrogen, and, less widely, sulphur. Other nutrients rarely need to be added at this stage. However, if cropping is prolonged and particularly if more highly demanding crops and cropping systems are introduced, the pattern of fertilizer response changes.

In the first place, phosphate deficiency intensifies in those soils in which it was not initially severe (Djokoto and Stephens, 1961a; Bromfield, 1967; Bouchy, 1970; Megie *et al.*, 1970), while the availability of nitrogen may temporarily improve (Djokoto and Stephens, 1961a). With both nitrogen and sulphur the greatest deficiencies occur in the years immediately after fallow, but while the narrowing of the C:N and C:S ratios under cultivation may subsequently increase availability, in most soils this increase is not sufficient to meet the demands of intensive crop production.

Secondly, new deficiencies appear, the most important being that of potassium. A gradual increase in the potassium requirement has been reported for cotton in Chad and Ivory Coast (Bouchy, 1970; Megie *et al.*, 1970) and for continuous groundnuts and groundnut-millet rotations in Senegal (Prevot and Ollagnier, 1959; Bockelee-Morvan, 1965a); and in Niger and Central African Republic deficiencies appeared after four and seven years' cropping, respectively (Lienart and Nabos, 1967; Braud, 1971). From Nigeria, Heathcote (1972a, b) reported substantial yield responses to potassium from maize, sorghum and groundnuts at sites at Kano, Samaru and Mokwa that had been cropped for a number of years. Responses were related to levels of soil exchangeable potassium (Table 53) and tended to increase as the experiment progressed. These results confirm the conclusions

TABLE 53

Crop responses to potassium at three sites in Nigeria, 1968-1971 (Heathcote, 1972a).

	Site		
	Kano	*Samaru*	*Mokwa*
Soil exch. K, me/100 g	0·09	0·20	0·06
*Mean crop responses to applied potassium, %:-			
Maize	60·3 (2)	7·7 (2)	41·1 (4)**
Groundnuts	22·8 (2)	1·4 (2)	-1·3 (2)
Sorghum	41·2 (3)	8·5 (2)	1·7 (2)
Cotton	—	0·2 (2)	—

*Figures in brackets indicate number of cropping years contributing to mean.
**Maize was grown in alternate years in two separate rotations. Potassium responses over the four years were 11·7% (1968), 22·9% (1969), 77·1% (1970) and 76·7% (1971).

of Wild (1971) that potassium deficiency is incipient in many savanna soils and that, in the first instance, levels of exchangeable K may be used to identify susceptible soils.

A benefit from liming has occasionally been demonstrated. Djokoto and Stephens (1961a) found an increasing lime response in continuously cropped soils in central and northern Ghana, but in combination with a negative lime x single superphosphate interaction this result suggested a deficiency of calcium. Interactions with potassium were also noted, but calcium appeared to be a more serious problem than potassium in many of these soils. In Nigeria, liming increased seed cotton yields at Samaru, but tended to interact with potassium at both Samaru and Kano (Heathcote and Stockinger, 1970). The appearance of positive lime x potassium interactions in both Ghana and Nigeria indicates that the calcium-potassium balance is a delicate one in many poorly buffered savanna soils and will need to be controlled under continuous cultivation.

It is often suggested that prolonged cropping will lead to micronutrient deficiencies, but so far few have been reported. Widespread boron deficiency in cotton has been identified (Braud *et al.*, 1969; Faulkner *et al.*, 1970), associated, in Cameroun at least, with intensified production and the use of NSP fertilizers (Fritz, 1971). In Nigeria symptoms are severest on drift (or loess) soils (Heathcote and Smithson, 1974).

Again in Nigeria, zinc deficiency has been found in maize at Mokwa, and molybdenum has increased yields of groundnut haulms at Samaru, in both cases in soils that had been cropped for a number of years (Heathcote, 1972a). As production intensifies, similar reports from other areas seem inevitable.

(c) *The use of fertilizers to maintain fertility*

The role of inorganic fertilizers in maintaining the fertility of savanna soils is as yet only supplementary to that of traditional methods. Moreover,

TABLE 54

Yields from long-term trials with inorganic fertilizers at Darou, Senegal*

Year	Trial 1: groundnuts only				Trial 2: groundnuts and sorghum			
	Haulm, kg/ha		Pods, kg/ha		G/nut pods, kg/ha		Sorghum grain, kg/ha	
	Control	With NPK**	Control	With NPK**	Control	With NPK***	Control	With NPK***
1	1401	2219	2296	2764	—	—	607	871
2	—	—	2122	2858	2061	2620	—	—
3	1051	1599	1571	1819	—	—	535	865
4	1336	2124	1371	2209	1845	2581	—	—
5	904	1416	1266	2004	—	—	424	773
6	747	971	933	1117	1766	2504	—	—
7	1046	1784	1360	2010	—	—	366	793
8	—	—	—	—	1770	2433	—	—

*Trial 1 from Prevot and Ollagnier (1959); trial 2 from Charreau and Tourte (1967).
**10:18:20, 10:13:20 and 10:13:20 kg/ha of N:P:K applied to the first, fourth and seventh years respectively of two separate groundnut-millet rotations.
***120 kg/ha of 6:20:10 and 100 kg/ha of 14:7:7 applied groundnut and sorghum crops, respectively.

this situation has been reflected in the sphere of research, for few long-term fertility trials have tested inorganic fertilizers in the absence of fallowing, manuring or the incorporation of green manures and residues, and none has been conclusive.

In the first of the two trials at Darou, Senegal (Table 54), low rates of NPK fertilizers, applied every third year to a continuous groundnut-millet rotation produced groundnut yields in the seventh year of cropping similar to those from first-year control plots. In the second trial, which involved annual applications of NPK fertilizers, groundnut and sorghum yields declined gradually over eight years, but remained well above those of the control plots.

At Yandev, Nigeria, plots receiving NP fertilizers once in each three-year rotation gave higher yields in the third rotation than did unfertilized plots following bush fallow (Table 55), and the increase was sufficient to pay for the fertilizer. However, neither here nor at Darou was it certain that the fertilizer maintained fertility. Rates of application were low, and the crop yield trend was, if anything, downwards.

Rather higher fertilizer rates have been used on certain long-term cotton trials. In Chad, on an alluvial soil, yields from 300 kg/ha NPS fertilizer every two years stabilized at about 3000 kg/ha, as compared with 1500 kg from control plots and 2500 kg from plots receiving biannual applications of farmyard manure (20 t/ha); but on sandy soils continuous cropping rapidly exhausted the soil and yields declined (Megie *et al.*, 1970).

During eleven years of continuous cotton cultivation at Bambari, Central African Republic, yields declined gradually on control plots, were

TABLE 55

Comparison of fertilized continuous cropping with alternate cropping and fallowing at Yandev, Nigeria (Watson and Goldsworthy, 1964).

		Crop yields, kg/ha	
Year	*Crop*	*Fertilized continuous cropping**	*Unfertilized fallow-cropping rotation*
1952	Yams	9420	7580
	Cowpeas	61	95
1953	Cereals	753	379
1954	Benniseed	393	321
1955-7		(Crops**)	(Fallow)
1958	Yams	7120	5930
	Cowpeas	16	28
1959	Cereals	1195	1028
1960	Benniseed	232	201

*167 kg/ha single superphosphate and 167 kg/ha ammonium sulphate applied to yams only.
**Same cropping cycle as 1952-4 and 1958-60.

TABLE 56

Yields of cotton under continuous cultivation at Bambari, Central African Republic (Richard, 1967).

Treatment	Mean seed cotton yields, kg/ha		
	First 3 yrs (1956-58)	*Last 3 yrs (1964-66)*	*All years (1956-66)*
Control	1294	1146	1178
Manure (20 t/ha/annum)	2068	2022	1909
NPK fertilizer*	1623	1919	1665
Manure + fertilizer	2079	2437	2024

*Fertilizer rates varied from year to year, but total application over eleven years was approximately 690 kg N, 340 kg P and 100 kg K per ha.

maintained on manured plots and increased gradually on plots receiving NPK fertilizers, but highest yields occurred where manure and fertilizer were applied together (Table 56).

Elsewhere in Central African Republic, at Grimari, yields of rotated crops have been maintained and even increased by the use of NPS fertilizer in conjunction with crop residue incorporation (Morel and Quantin, 1972). It was concluded that fertility could be maintained for at least ten years using mineral fertilizers, but only after the physical state of the surface soil had been improved by the appropriate cultivation techniques. Once this had been done, good crop growth was itself a factor in the improvement and maintenance of fertility, through the increased return of fresh organic matter.

The view that the maintenance of fertility is intimately related to the physical state of the soil and the rate of turnover of organic matter is a widely held one, at least among francophone workers, and investigations of inorganic fertilizers which exclude consideration of interactions with the management of soil and crop residues are likely to have only limited success in predicting and overcoming the problems facing the ordinary farmer. Thus the question frequently asked, whether the fertility of savanna soils can be permanently maintained with inorganic fertilizers alone, is an unrealistic one, if it ignores the management aspect. All practical agricultural systems, whether mixed or purely arable, give rise to organic waste products, manure or crop residues which, where they have no other use, may be returned to the soil. In addition to the improvement of physical and physico-chemical properties, the return of these materials conserves plant nutrients, and the conservation of nutrients within the soil-plant system is fundamental to an economic fertilizer policy. Even where incorporation or mulching is impracticable, the burning of residues on the field at least ensures the return of a proportion of their ash content.

Many of the most relevant long-term fertilizer experiments, therefore, have been those testing inorganic fertilizers in combination with different rotation, fallowing and manuring treatments, as at Grimari (Morel and Quantin, 1972). The results, though difficult to interpret with certainty, tend to reinforce the view that fertilization should be organized not on an annual *ad hoc* basis but according to the needs of the whole rotation.

The value of rotation fertilization was demonstrated by Tourte *et al.* (1967), in Senegal. They compared annual fertilization with a fertilizer programme comprising:

(i) basal phosphate: 500 kg ground rock phosphate per ha, applied to the fallow and incorporated with green manure,

(ii) 50 kg K/ha on each groundnut crop, and

(iii) 60 kg N/ha on the cereal crop,

on a fallow-groundnut-cereal-groundnut rotation. The input of fertilizer was higher under rotation fertilization, but the much greater yield response made it the more profitable system. This approach assumes a completely settled system of "semi-intensive" agriculture, with animal traction and fixed four to six year rotations which include one or two years of herbaceous fallow. Animal traction permits deep cultivation and incorporation of unburned fallow vegetation, manure and crop residues, techniques regarded as essential to the success of the system.

Intrinsic to this work are the ideas of basal and maintenance fertilization (Chaminade, 1970) and the practice of green manuring or fallow incorporation. These are now discussed briefly below.

(i) *Basal fertilization*

Basal (or correction or redressment) fertilization aims to raise substantially the basic fertility of the soil with one large dressing, its size determined by the degree of the existing soil deficiency. Because of the widespread occurrence of phosphate deficiency, this has usually meant heavy phosphate applications using local rock phosphates. Workable deposits of rock phosphates occur in Senegal, Mali, Niger and Togo, and these have been widely tested as cheap sources of fertilizer phosphate. The main factors determining their effectiveness appear to be rainfall and method of application.

As mentioned already, crop response to rock phosphate increases with increasing rainfall. Experiments in the 1950s over a wide range of sites revealed a highly significant correlation between rainfall and groundnut response to a basal dressing of 500 kg/ha rock phosphate applied to the fallow the previous season (C.R.A., Bambey, 1957). A zero response at 550 mm rainfall increased to one of about 25 per cent at 1150 mm; and similar trends were seen in the two subsequent cropping seasons. Further work has confirmed this result and led to the recommendation that rock phosphate be used only where mean annual rainfall exceeds 700—800 mm.

The optimum size of basal dressing depends on soil type. Lienart and Nabos (1967) recommended 33 kg P/ha for dune soils at Tarna, Niger, and Jenny (1965) found 22 kg P/ha enough for upland sites in Upper Volta. In their review, Pichot and Roche (1972) gave data for more than twenty upland sites in West Africa. The phosphate required to correct deficiency was, at all of them, in the range 20-55 kg P/ha. However, on hydromorphic and lowland alluvial soils the requirement is more variable. On a hydromorphic rice soil at Djibelor, Senegal, 44 kg P/ha was sufficient, but an acid sulphate soil needed 130—175 kg P/ha (Pichot and Roche, 1972). On very strongly phosphate-adsorbing soils, as in the Senegal valley, applications of more than 400 kg P/ha may be necessary (Couey *et al.*, 1965; 1967).

One advantage of the use of calcium rock phosphate is the contribution it makes to the calcium status of the soil. Poulain (1967) quoted results from a rotation in which the pH of the control plot fell 0·3 units over 4 years, while that of a plot that received rock phosphate at 1000 kg/ha remained constant.

(ii) *Maintenance fertilization*

The purpose of maintenance dressings is to balance nutrient losses through crop removal, leaching and erosion, in order to maintain the fertility level established by the basal fertilizer; but in practice this approach is severely handicapped by a lack of basic information. Losses through leaching and erosion are known for only one or two sites in the whole region. Crop removal data are easier to obtain, but few have yet been published. Those given in Table 57 are far from complete and are not related to yield level;

TABLE 57

Nutrient removal in economic yield, kg per tonne of crop

	*Ref**	*N*	*P*	*K*	*Ca*	*Mg*	*S*
Groundnuts							
pods	1	—	4	—	—	—	—
	5	50	4	9	0·5	2	2·6
	6	30	2·7	—	—	—	1·8
haulm	5	36	2	35	18	10	4
	6	15	1·2	—	—	—	1·6
pods and haulm	7	51	4	20	14	—	—
Millet grain	5	26	3	10	0·6	1·6	2·8
Sorghum grain	1	—	2	—	—	—	—
	2	21	2·4	4·1	0·8	2·7	0·5
	3	19	2·4	3·5	0·5	1·9	0·8
	7	22	2·2	4·2	1·0	—	—
Maize grain	3	9·3	1·9	2·5	0·4	0·7	—
	4	13	3	4	0·1	1·0	—
Cotton seed and lint	7	28	4·8	10	2·1	—	—

*1, Goldsworthy and Meredith, 1959; 2, Jacquinot, 1964; 3, Dupont de Dinechin, 1967a; 4, Jones and Stockinger (unpublished data, Samaru); 5, Bertrand *et al.*, 1972; 6, Bromfield, 1973; 7, Tourte, 1971.

TABLE 58

Nutrient balance for a five-year rotation at Sefa, Senegal (Charreau and Fauck, 1970).

Crop	Assumed yield, kg/ha	Treatment of residue	(a) Calculation for maintenance dressings				(b) Calculation from optimal response in fertilizer field trials			
			Nutrient removal, kg/ha				Nutrient applied, kg/ha			
			N	*P*	*K*	*Ca*	*N*	*P*	*K*	*Ca*
Fallow	10,000	Incorporated	0	0	0	0	40	122	—	279
Maize	5,000	Incorporated	95	22	50	4	120	—	75	—
Rice	4,000	Removed	86	19	101	26	69	—	—	—
Groundnuts	3,000	Removed	0	10	54	39	10	—	50	—
Millet	3,000	Burnt on field	120	13	27	6	69	—	42	—
Five year total:			301	64	232	75	308	122	167	279
Cumulative leaching loss (5 years):			150	0	83	536				
Total loss:			451	64	315	611				
*Mean annual loss or addition:			90	13	63	122	61	24	33	56

*Comparison of data in this line shows that fertilizer dressings developed from the results of field trials supply more P, but less N, K and Ca, than is lost from the soil by crop removal and leaching.

much more detailed information is needed before reliable general estimates of maintenance dressings can be made.

Charreau and Fauck (1970) calculated maintenance dressings for a fallow-maize-rice-groundnuts-millet rotation at Sefa. They calculated annual leaching losses by extrapolating Bambey lysimeter results to existing chemical analyses of Sefa soils and estimated crop removal, assuming high yields, from known plant analysis data. From these figures, total nutrient loss and hence the required maintenance dressings could be calculated (Table 58a).

Charreau and Fauck compared these calculated maintenance dressings with fertilizer recommendations based on the results of field trials (Table 58b). For N, K and Ca, these empirical values were lower than the theoretical maintenance values. For N the difference was not large and could in part be attributed to contributions from rainfall and biological fixation, which had been ignored in the original estimates. In any case, a shortage of nitrogen fertilizer, if real, would not be likely to deplete the soil but to lead to a poor crop and, eventually, an upward revision of the fertilizer recommendations. However, for potassium and particularly calcium the unfavourable balance was more serious. As other work at Sefa has shown, the cumulative loss of calcium and the consequent drop in pH tends to have little effect on crops until a low threshold value is reached, perhaps around pH 4·5. Thereafter the depression of yields can be very great.

The occasional lime application that these authors advocate for Sefa soils will usually be necessary wherever annually leached soils are cultivated without recourse to long fallows. Amounts will vary widely according to the severity of leaching and the fertilizer regime. Acidifying fertilizers like ammonium sulphate and urea will increase the requirement.

One may conclude that fertilizer recommendations based solely on the results of short-term trials are liable to fail in continuous agriculture, unless they are continually revised to allow for long-term changes in the soil. The maintenance principle of replacing all lost nutrients is a sound one, particularly for poorly buffered soils, though with phosphate more will generally be needed to allow for its reaction with the soil. Its large-scale implementation must, however, await a better understanding of the nutrient balance-sheet, especially as regards leaching losses.

(iii) *Green manuring*

This practice involves incorporating the green material from one season's plant growth into the soil just before the end of the rains. Natural herbaceous fallow may be used, in which case the ploughing-in may be thought of as an alternative to burning for clearing the fallow. Alternatively, a cultivated species, cereal or legume, may be planted, fertilized and then incorporated before reaching maturity.

The practical drawbacks are readily apparent, not least to the traditional farmer. The initial labour of planting produces no harvest, and much effort

TABLE 59

Effect of different forms of soil regeneration on crop yields and leaf content of potassium at Darou, Senegal (Delbosc, 1967).

Form of Regeneration	*1st crop: groundnuts* Mean yield, kg/ha		*1st crop: groundnuts* Leaf K, %		*2nd crop: sorghum* kg/ha		*3rd crop: groundnuts* kg/ha	
	Fert'd	*Unfert'd*	*Fert'd*	*Unfert'd*	*Fert'd*	*Unfert'd*	*Fert'd*	*Unfert'd*
	*	n.s.	n.d.	n.d.	**	*	*	**
Green manure	2478	1964	0·89	0·86	443	337	2204	1561
Burned fallow	2702	2260	1·37	1·43	846	566	2426	1693

Significant differences between forms of regeneration: * 0·05 level, ** 0·01 level.
Experiment set up in 1954 on Ferruginous Soil. Data are means from period 1959-66.

is needed to work the bulky plant material into the soil. For this, some form of traction is almost essential.

There is little evidence that green manuring increases soil organic matter, except ephemerally, for the material incorporated is largely unlignified, and decomposition is very rapid. It may, however, have a small, less transient, effect on the distribution of humic fractions within the soil organic matter (Bouyer, 1959).

The argument in favour of green manuring is that rapid decomposition releases accumulated nutrients in readily available form to the subsequent crop (Poulain, 1965), but this is difficult to demonstrate convincingly. Delbosc (1967) described an experiment at Darou, Senegal, in which a groundnut/millet/groundnut rotation followed either two years' natural fallow, burnt, or a one-year cereal green manure, incorporated. Yields in all three cropping years were superior after burnt fallow, and the data for groundnut leaf K-content (Table 59) suggest one reason for it. The green manure had not been effective in mobilizing and releasing potassium to the following crop. This author concluded that, despite a reduction in the time the land was out of production, green manuring offered little advantage, especially in view of the practical difficulties involved.

Others have reached similar conclusions. In Central African Republic, Morel and Quantin (1972) observed no significant improvement in soil physico-chemical characteristics from green-manuring and pointed out the risk to surface structure from the heavy equipment used to incorporate the bulky material.

Nevertheless, a strong case can be made out for green manuring in those parts of the western savanna where the rainy season is very short and its onset sudden (Poulain, 1965). In Senegal deep cultivations have proved beneficial (Charreau and Nicou, 1971), but it is difficult to find time for them. Ploughing at the beginning of the season delays planting, and delayed planting reduces yields. Ploughing at the end of the season is impossible except after a short-season crop. The inclusion in the rotation of a brief fallow or a green-manure crop, with incorporation just before the end of the rains, permits at least one such deep cultivation per rotation. The incorporated material improves the structure of these very sandy soils (Charreau and Nicou, 1971), and, furthermore, ploughing in with the green manure has proved an effective way of applying basal phosphatic fertilizers (Poulain and Mara, 1965; Dupont de Dinechin, 1967b). We shall return to this subject in the next chapter.

3 FORMS OF FERTILIZER

The most widely used fertilizers in the savanna have until recently been single superphosphate (8 per cent P) and ammonium sulphate (20·5 per cent N). The present trends are:

(i) towards higher analysis materials to save transport costs;
(ii) away from ammonium sulphate, to reduce soil acidification;
(iii) towards compounds and mixed fertilizers, to meet more closely the needs of the plant, while limiting the complexity and labour of the spreading operation; and
(iv) towards rock phosphate, where readily available, as a basal fertilizer.

Much of the phosphate is applied to groundnuts. Where the source is single superphosphate its sulphur content (12 per cent) is also useful, although even for this crop the proportion of sulphur is unnecessarily high, as is implied by the results of Bromfield (1973). However, a change to the higher-analysis triple superphosphate (20 per cent P) would mean that in many groundnut areas sulphur would have to be applied separately as elemental S, gypsum or, possibly, ammonium sulphate. Ammonium phosphate has been found suitable for cereal crops (Goldsworthy, 1967c), as has potassium metaphosphate, at least in small amounts, for sorghum (Dupont de Dinechin, 1967c).

Of the nitrogen fertilizers, the most concentrated, anhydrous ammonia and aqua ammonia, are not practicable under present farming conditions. In northern Nigeria, urea (46 per cent N) and calcium ammonium nitrate (20-26 per cent N) have given equal or only slightly smaller crop responses per unit of nitrogen than ammonium sulphate (Goldsworthy, 1967c; Jones, 1975b). Urea is cheaper per unit of nitrogen than ammonium sulphate or calcium ammonium nitrate. Ammonium phosphate, ammonium polyphosphate and certain urea derivatives are either competitive in price or might become so.

However, the costing of nitrogen fertilizers should also take into account the cost of the lime needed to neutralize the acidity they cause. Urea, ammonium salts, and compounds and mixtures containing them, are all acidifying in some degree; and lime is neither common nor cheap in the savanna. For this reason the safest fertilizer that can be recommended at present for general use is calcium ammonium nitrate, in which the acidifying effect of the ammonium nitrate (approx 70 per cent by weight of the mixed fertilizer) is largely neutralized by its content of calcium carbonate. Lower analysis grades (20 per cent N) are completely neutral. This material may be replaced by urea or other sources where the soils are neutral and well-buffered. Sulphur-coated urea and other slow-acting nitrogen sources may find application in more humid areas when their cost has been reduced.

As has been pointed out, where cultivation is intensive, S, K, Mg and B and perhaps Ca, Mo and Zn may also be required. The eventual need is likely to be for a range of multinutrient materials, which will enable farmers to apply in one operation a complete maintenance dressing for the particular crop being grown. But a great deal more information is required before the specifications for such materials can be confidently drawn up.

Chapter 11

THE CONTROL OF SOIL PHYSICAL PROPERTIES

The emphasis in Chapter 10 was on the maintenance of the nutrient status of the soil; but for optimum crop growth the soil must be in a physical condition that permits the efficient utilization of its nutritional potential. No fertilizer regime can be said to be maintaining fertility if the poor physical state of the soil limits root growth, hampers infiltration and leads to erosion. Further, water supply is a major factor limiting growth. This chapter therefore reviews experimental work on soil management, firstly, erosion control and the management of rainwater, and secondly, the relationships between soil cultivations, crop rooting and yields.

1 THE CONSERVATION OF SOIL AND WATER

The primary causes of erosion were discussed in Chapter 4. Raindrop impact causes soil splash, it compacts the soil surface, thereby reducing infiltration, and it disperses soil particles, facilitating their removal by flowing water. At the same time surface flow, particularly if it is constricted into fast-moving streams, scours soil from along its path. In this section we examine the effectiveness of soil and crop management techniques for erosion control.

(a) *Raindrop impact and soil cover*

The destructive effect of raindrop impact is minimised by the rapid establishment of a protective crop cover. Fournier (1967) emphasized the importance of early planting. He quoted results of Hudson (1957) in Rhodesia, where late planting of tobacco had increased annual soil loss threefold relative to early planting (from 4 to 11·2 t/ha), although runoff remained the same (15 per cent of 845 mm rain). At Sefa, a 15-day delay in planting groundnuts doubled erosion:

Planting date	*Erosion, t/ha*
18 June	3·0
2 July	6·4

(Charreau and Fauck, 1970).

Planting date is not the only factor. The rate of establishment of crop cover depends also on fertilizer regime and crop density. Further results from Rhodesia (Hudson, 1957; Hudson and Jackson, 1959) showed that erosion under heavily fertilized maize planted at high density was less than half that under a moderately fertilized crop at moderate density (Table 60).

TABLE 60

Erosion from maize plots at different productivity levels at Mazoe, Rhodesia (Hudson and Jackson, 1959).

	*Productivity level**			
	1		2	
Season	*Run off,* %	*Erosion,* *t/ha*	*Run off,* %	*Erosion,* *t/ha*
1955-6	8	6·3	14	22·6
1956-7	18	26·4	20	58·8
1957-8	1	1·5	4	4·5

*1. 37,000 plants/ha; heavy fertilization.
2. 24,700 plants/ha; moderate fertilization.

As this work shows, early planting combined with high crop densities and high rates of fertilization can be a very effective method of erosion control; but it should be applied with caution. Early planting at high fertility levels increases the risk of crop loss if the early rains are followed by drought, and erosion from fields carrying a failed crop can be very high (Table 20).

Protection of the soil surface may also be achieved by mulching. In an extreme case, that of mulched bananas under high rainfall at Adiopodoume, Ivory Coast, annual soil loss was only 5 kg/ha, which was less than that under forest cover (Fournier, 1967). In the savanna too, favourable effects of mulching on crop yields have frequently been reported (Nye, 1952b; Lawes, 1957; Djokoto and Stevens, 1961a, b; Ashrif and Thornton, 1965). In all cases, the advantage was attributed to physical effects, principally improved rainfall acceptance, reduced runoff and surface capping, and improved moisture conditions and aeration in the topsoil.

Mulching studies at Ibadan, Nigeria, have shown that direct seeding of maize into soil covered with partially decomposed crop residues increases yields and keeps soil losses low. The advantage to the crop from mulching appears to lie in lower soil temperatures at the seedling stage, suppression of weed growth and improved moisture storage in the root zone (Lal, 1973). These are all factors one would expect to be equally important in the savanna.

The main problems with mulching are the risk of a carry-over of insect pests, the lack of availability of suitable material and, very often, the uneconomic nature of the operation. Both the labour of spreading and the cost of nitrogen fertilizers needed to overcome immobilization can make the technique unprofitable despite its favourable effect on crop yield and soil conservation. It may be easier in a mechanized system. In Rhodesia a trash-mulching technique, in which maize residues were chopped and partially incorporated *in situ,* reduced water and soil losses by nearly 50 per cent (Hudson 1957; Hudson and Jackson, 1959). A fuller discussion has been given by Hudson (1971).

(b) *Control of runoff and erosion*

Runoff is both an agent of erosion and the medium of transport carrying the eroded material away. Control of runoff is therefore essential to control of erosion. A clear account of the available techniques has been given by Hudson (1971), and critical assessments of the use of many of them under savanna conditions have been made by Fournier (1967) and Roose (1967). The following account is based largely on their findings and on data (summarized in Table 61) from runoff plots at six savanna sites.

In the experimental work on runoff plots three different types of comparisons may be distinguished: first, between the effects of different forms of surface micro-relief (e.g. ridges versus flat cultivation); secondly, between complete cultivation systems, involving different ploughing and planting regimes; and thirdly, between larger-scale conservation measures, such as contour banks and buffer strips. They are here considered separately, with a short final discussion of field-scale conservation measures not tested by runoff plots.

(i) *Ridge versus flat cultivation*

In six of the ten sets of data in Table 61 (A, B, C, D, E, F) comparisons were made between different types of soil cultivation as they affected the microrelief of the field. Although there were considerable differences in soils, slopes and type of cover between sites, a number of generalizations may be made:

1. Ridges running down the slope always encourage erosion. By concentrating the flow of water into relatively narrow channels following the line of greatest gradient, runoff velocity and hence the scouring effect are maximized.

2. Similar remarks apply to mounds, for whatever the orientation of the rows of the mounds, there is little obstruction to down-slope flow in the channels in between; but mulching and cross-tying of mounds were effective in preventing erosion at Bouake.

3. Contour ridges, tied or untied, give little runoff or erosion, at least under experimental conditions.

4. Results from flat cultivation are very variable. At Samaru and, less convincingly, at Bouake flat cultivation was less erosive than ridges down the slope, but not at Niangoloko. At Niangoloko, Sefa and Boukombe, flat cultivation led to more erosion than contour ridges.

The above generalizations require qualification and comment. In the first place, it is necessary to distinguish between measures which control erosion by preventing runoff (tied contour ridges, tied mounds) and those which are designed only to reduce its rate of flow (flat cultivation and graded ridges, i.e. ridges aligned slightly off the contour). Over much of the savanna there is a humid period during which rainfall markedly exceeds evapotranspiration. Measures to prevent runoff at this time may not only be difficult to implement but also harmful, encouraging waterlogging and the loss of

TABLE 61

Runoff and erosion data from runoff plots at six savanna sites.

Ref.	*Site*	*Soil type*	*Plot dimensions: Length, m*	*Slope, %*	*Results are means of:— Reps.*	*Years*	*Management*	*Cover*	*Mean rainfall, mm.*	*Runoff, %*	*Erosion, t/ha/annum*
A.	Samaru (Nigeria)	Leached Ferruginous	183	0·3	1	4/5	a) Flat cultivation	Bare fallow	1062	25·2	3·8
					1	4/5	b) Flat cultivation	Arable crops	1062	20·3	4·0
					1	4/5	c) Ridges down slope	Arable crops	1062	27·9	19·6
					1	4/5	d) Alternate tied ridges	Arable crops	1062	13·2	5·7
					1	4/5	e) Broadlands	Arable crops	1062	20·4	21·0
B.	Bouake (Iv. Coast)	Ferrallitic	50	4·0	1	2	a) Traditional mounds	Yams	1105	12·3	2·7
					1	2	b) Mounds, tied and mulched	Yams	1105	0·0	0·0
					1	2	c) Ridges down slope	Maize/cotton	1045	13·9	14·6
					1	3	d) Tied contour ridges	Maize/cotton	1118	0·0	0·0
					1	2	e) Flat cultivation	Upland rice	1205	1·9	0·3
					2	1	f) Flat cultivation	G'nuts/tobacco	1145	12·3	12·4
					1	1	g) Flat cultivation	G'nuts/fallow	945	0·0	0·0
C.	Niangoloko (Upper Volta)	Leached Ferruginous	50	0·5	2	4	a) Ridges down slope	G'nuts-millet	1137	2·3	1·6
					1	4	b) Ridges on contour	G'nuts-millet	1137	0·4	0·4
					2	4	c) Flat cultivation	G'nuts-millet	1137	7·5	6·4
D.	Niangoloko (Upper Volta)	Leached Ferruginous	50	2·5	?	4	a) Closed contour ridges	G'nuts-millet	1137?	0·9	1·4
					?	4	b) Open graded ridges (0·5%)	G'nuts-millet	1137?	6·3	6·1

					?	4	c) Flat cultivation	G'nuts-millet	1137?	12·2	13·2
E.	Sefa (Senegal)	Leached Ferruginous	50	1·0	?	?	a) Flat cultivation	Rice	?	22·3	6·2
					?	?	b) Flat cultivation	Groundnuts	?	9·1	3·1
					?	?	c) Tied contour ridges	Cotton	?	0·6	0·5
F.	Boukombe (Dahomey)	Leached Ferruginous	42	3·7	?	1	a) Flat cultivation	Millet	875	11·7	1·3
					?	1	b) Open contour ridges	Millet	875	2·0	0·2
G.	Sefa (Senegal)	Leached Ferruginous	40	2·0	1	3	a) Control	Maize x 2; rice	1123	31·2	20·8
					1	3	b) IRAT system	Maize x 2; rice	1123	25·8	8·3
			50	1·5	1	3	c) Control	Maize x 2; rice	1123	21·3	13·6
					1	3	d) IRAT system	Maize x 2; rice	1123	14·9	3·3
			50	1·0	2	3	e) Control	Maize x 2; rice	1123	24·2	5·9
					2	3	f) IRAT system	Maize x 2; rice	1123	15·6	2·6
H.	Sefa (Senegal)	Leached Ferruginous	40—50	1·5—2·0	1/2	8	a) Traditional cultivation	Arable crops	1180	25·9	7·2
					1/2	8	b) Mechanized cultivation	Arable crops	1180	33·0	14·1
I.	Bouake (Ivory Coast)	Ferrallitic	46	4·0	1	2	a) Control	G'nuts; maize	1180	12·4	7·6
					2	2	b) 2m. grass buffer strip	G'nuts; maize	1180	5·0	0·9
					2	2	c) 4m. grass buffer strip	G'nuts; maize	1180	3·7	0·6
J.	Allokoto (Niger)	Lithomorphic Vertisol	70	3·0	1	3	a) Traditional cultivation	Sorghum; cotton	452	16·3	8·6
					1	3	b) Stone contour banks	Sorghum; cotton	452	3·9	0·7
					1	3	c) Contour grass buffer strips	Sorghum; cotton	452	4·1	1·5
					1	2	d) Grass contour banks and ditch	Sorghum; cotton	452	0·3	0·1

Sources and notes, p. 176

Sources and notes (Table 61):—

A. Kowal (1970b, c); crop rotation, 1964-1968: cotton-cotton-sorghum-g/nuts-cotton. Runoff data for all five years; erosion data for 1965-68 only. For exposition of "broadlands" (e) see Hill (1961); at Samaru alternate ridges in broadlands were tied, the broadlands ridges running down slope.

B. Berger (1964); two crops a year, hence maize/cotton, etc. See also Bertrand (1967a).

C, D. Christoi (1966), Fournier (1967), respectively. Although these come from the same site over approximately the same period of time, treatments, slopes and results all differ, so it is assumed here that there were two separate experiments. In C, groundnuts and millet were grown in rotation with 2 or 3 fallow years, but only cropped years are reported.

E. Fournier (1967); no replication or rainfall data given.

F. Verney and Williame (1965); flat-cultivated plot was ploughed to 20 cm, and this was thought to have increased soil loss.

G. Charreau (1969). Control: disc harrow to 5-10 cm in wet soil at beginning of season, planting 1st July.
IRAT system: mouldboard plough to 25-30 cm at end of previous season and late dry-season harrowing; planting 10-15th June.

H. Roose (1967). Mechanized cultivation in this case was a shallow one at the beginning of the rains, as in G: control treatment.

I. Roose and Bertrand (1971); in the second of the two years crops were planted on ridges down the slope, but buffer strips still effective.

J. Roose and Bertrand (1971); all treatments apart from (a) were ploughed and contour ridged. See also Birot and Galabert (1967).

nutrients by leaching. Under these conditions management systems which allow a slow non-erosive runoff of surplus water are likely to be more satisfactory.

A second point concerns the ease with which systems tested at plot level can be transferred to the open field. Open contour ridges gave good control of erosion on narrow experimental plots and reduced runoff to very low levels at both Niangoloko and Boukombe; but the construction of ridges across a wide field precisely on the contour is a difficult operation. Only a small variation in level can cause water to gather at one point in a furrow, where it overtops during a storm and initiates ridge collapse and rill erosion down the length of the field. Tying ridges, by creating a series of small basins along each furrow, largely overcomes this problem, and tied ridges can take all but the largest storms without overtopping.

Hudson (1971) emphasized the necessity of two safety devices for tied ridges. First, the ties should be lower than the ridges and the ridges themselves should be graded, so that any sudden release of runoff water is along the furrow and not down the slope. Secondly, there should be a support system of conventional contour terraces to cope with runoff from exceptional storms.

However, if there is no requirement to conserve water, then untied graded ridges can be used to control runoff without risk of overtopping, waterlogging or excessive leaching; but few results have been found that demonstrate this. At Niangoloko, ridges graded at 0·5 per cent across a slope of 2·5 per cent lost the equivalent of 6 t soil/ha/annum, which was less than half that lost from flat cultivation. These graded ridges were short, because of limited plot width, but at Samaru, Kowal (1970b, c) had ridges of 0·3 per cent gradient running across a slope of 1·5 per cent for a distance of 183 m. Runoff and erosion, collected at the ends of the ridges, were 27·9 per cent of the rainfall and 19·6 t/ha/annum, respectively, values high enough to raise doubts about the safety of long, graded ridges. Gentler gradients would be safer. Since runoff velocity, and hence erosion, is related to slope, the ideal ridge gradient is one which is only just sufficient to drain off excess water.

The alternative to ridges is flat cultivation, which under ideal conditions permits slow non-erosive sheet flow .One advantage of flat cultivation is that, unlike ridges, it imposes no limitation on planting geometry. Crops may therefore be sown at spacings which ensure optimum utilization of radiation and soil nutrients and also, what is important for soil conservation, the maximum soil cover at all stages of growth (Kowal, 1970b, c).

However, a clean-weeded, flat-planted crop offers little resistance to the movement of water, and any excess flows off almost immediately. This means that during heavy storms the sheet of water becomes rapidly deeper down the slope and, encouraged by minor variations in surface relief and old lines of natural drainage, often tends to concentrate into streams. Rills and gullies are rapidly formed, and loss of soil and crop may locally be severe. It is doubtful whether the erosion data for flat-planted crops in Table 61 sufficiently allow for this phenomenon, for the carefully prepared surface and the small catchment area of a runoff plot do not provide the conditions for serious rill and gully formation. Actual erosion from large flat-planted fields may therefore be much greater than these figures suggest.

Field dimensions are critical, because if fields are laid out as graded contour terraces and are separated by adequate banks, then rill formation may be of minor importance. Kowal (1970d) has discussed terrace width in the context of one exceptionally heavy storm at Samaru. He observed that on one terrace laid across a slope of 1·4 per cent, rills started about 35 m from the top of the slope and became serious at about 55 m. In this situation the limitation of terrace width to, say, 50 m might therefore be expected to keep the problem of rill formation within reasonable bounds. In general, the safe width will vary with slope and soil type (Hudson, 1971).

(ii) *Cultivation systems*

Mechanized cultivation was reported to increase runoff and erosion relative to traditional hand cultivation both at Sefa (Roose, 1967, Table, Table 61, H) and at Boukombe, Dahomey (Verney and Williame, 1965). This was probably because frequent discing and harrowing with heavy equipment destroyed surface structure more rapidly than hoeing. However, deep ploughing (to 20—30 cm) at the end of the season appears to avoid this, as well as permitting earlier planting the next year. Charreau (1969) compared the effect of the two systems on runoff and erosion:

1. Control: disc ploughing of wet soil to 5—10 cm depth at the beginning of the wet season; mean planting date, 1st July.
2. IRAT system: mouldboard ploughing to 25—30 cm depth at the end of the wet season with one harrowing operation at the end of the dry season; planting date: 10—15th June.

Runoff and erosion were lower under the IRAT system (Table 61, G). Charreau suggested two main reasons: deep ploughing had improved the porosity and infiltration, and ploughing at the end of the wet season had allowed earlier planting and so an earlier establishment of crop cover. The importance of earlier crop establishment was indicated by the following data for the runoff water:

	Turbidity, g/litre		
System	1966	1967	1968
Control	9·25	3·17	1·05
IRAT	6·26	1·38	1·02

The lower turbidity accounted for the greater reduction of erosion relative to runoff under the IRAT system.

(iii) *Other conservation techniques*

The techniques discussed so far have all been concerned with the nature of the preparation of the surface soil; but there are other approaches to erosion control. One is strip cropping, with strips of different crops, planted on the contour, alternating down the slope. Clearly, the success of this system depends on the availability of at least one "non-erosive" crop able to exert a braking effect on flowing water. A crop like groundnuts, with plenty of foliage near the ground, should be effective. Unfortunately, in an experiment to test strip-cropping at Sefa a systematic error in design obscured any benefit there might have been in using strips of groundnuts to control erosion from other crops (Roose, 1967; Fournier, 1967).

An alternative approach is to plant narrow "buffer-strips" of grass on the contour between wider strips of cropped land. This was tried at Bouake, Ivory Coast, and at Allokoto, Niger, by Roose and Bertrand (1971). At Bouake (Table 61, I) a 2 m strip of grass at the bottom of a plot 46m long and of 4 per cent slope was sufficient to halve runoff and reduce erosion losses to very low levels, even if, as in the second experimental year, crops were planted on ridges down the slope. At Allokoto losses from plots with buffer strips of *Andropogon gayanus*, planted at an 80 cm vertical interval on a 3 per cent slope (i.e. approximately 27 m apart), were less than one fifth those from traditional cultivation, although in this experiment conservation was assisted by contour ridging in all plots except that carrying traditional cultivation.

Although they have not been widely tested, it seems probable that grass buffer strips could prove a useful control measure, but there is need for caution in their application. First, it is essential that the strips themselves are completely protected from erosion by a sufficiently dense plant cover; problems of establishment may arise where the season is short, as at Allokoto (Birot and Galabert, 1967). Secondly, the grasses should not be rhizomatous or they may spread to cropped land and be difficult to eradicate. Thirdly, since absorption of water by the grass strips is usually not complete, there is the risk of a dangerous build-up of runoff towards the lower end of the field (Hudson, 1971). Buffer strips should not therefore be regarded as an alternative to graded terraces, but rather as a supplementary measure, which might enable terrace widths to be increased and rill erosion to be controlled where the terrace surface is uneven.

(iv) *Terraces and larger-scale conservation measures*

Two other treatments in the Allokoto experiment (Table 61, J) were stone contour banks and grassed contour banks with ditches, both of which gave greater erosion control than grass buffer strips. These measures effectively turned each plot into a number of absorption terraces or basins from which runoff could occur only in exceptional circumstances. Thus water was conserved as well as soil.

In wetter areas, closed terraces of this type would produce waterlogging and overtopping from one terrace to the next. It is necessary in these areas to align the banks between terraces at a small angle to the contour and to allow slow runoff along shallow grassed ditches at the foot of each terrace. The situation is in some ways analogous to that of the graded ridges described above, but the scale is larger and the grass of the ditches confers a resistance to erosion and traps much of the soil eroded from the cultivated area of the terrace.

The general principles of laying out graded contour terraces of this type were described in some detail for Rhodesian soils by Aylen (1949) and Cormack (1951). No comparable publications have been found for West Africa. Palmer (1958) described the terrace design for the Niger Agricultural Project at Mokwa, Nigeria, based on Cormack's figures: on slopes of 1—3 per cent and up to 4·8 km long, cultivated terraces were separated by shallow grassed ditches and grassed earth banks constructed at approximately 1·5 m vertical intervals down the slope. The maximum length of terrace was 550 m, and the lateral slope of its drainage ditch increased towards the outlet to cope with the disposal of an increasing volume of water, as follows:

Distance from start of terrace, m	*Ditch slope, %*
0—275	0·1
275—365	0·2
365—455	0·3
455—550	0·4

Each ditch drained into a grassed watercourse, the width of which increased down the slope to accommodate the increasing volume of water.

The experimental farm at Samaru is of similar design. Terraces up to 365 m long are laid out across a topographical slope of about 1·5 per cent at vertical intervals of just over one metre, their gradient increasing from 0·2 per cent at the top end to 0·4 per cent at their outlet into grassed watercourses. This design has successfully controlled erosion for more than ten years and was effective in removing a high volume of runoff from flat-planted crops without excessive soil loss or damage to watercourses during an exceptional storm of 71 mm in 1969 (Kowal, 1970d).

Measures of this type, however, are suitable for the small farmer only where conservation schemes are planned for whole catchment areas, and where equipment is available for constructing terraces and protective earth banks. For the farmer without this support, and with only hand implements or animal traction, the best technique is probably planting on ridges aligned close to the contour. Buffer strips of grass between fields and the careful siting and protection of footpaths, which often act as stream beds during heavy rain, are essential where long slopes are under cultivation.

(c) *Conservation of water*

Crop production is limited by lack of water for at least part of the year in all areas of the savanna. Management methods must therefore be sought which make most efficient use of the existing rainfall (Russell, 1968b). However, where the humid period is prolonged, efficient use of rainfall does not necessarily mean the prevention of all runoff, for there is a limit to the soil's storage capacity. A balance has to be struck between conservation of soil and water by runoff control and the avoidance of surface waterlogging and excessive leaching. Factors in this balance include annual rainfall and its distribution, soil type (infiltration and water-holding characteristics), and crop (planting date, length of growing season, pattern of water use). So little is yet known of these factors that the choice of conservation system is still largely a matter of guesswork; hence the importance of the recently initiated interest in agroclimatological studies in the savanna (Cocheme and Franquin, 1967 a, b; Charoy, 1971; Kowal and Knabe, 1972; Kowal and Andrews, 1973; Kowal and Faulkner, 1975).

In more arid areas the choice is clear. All rainfall must be retained, and techniques which reduce runoff, improve infiltration and increase water-holding capacity of the soil are likely to benefit crop growth. Data from the first year of the runoff plots at Allokoto illustrate the high returns that may be obtained from reducing runoff at a semi-arid site (Birot and Galabert, 1967). In 1966 (total rainfall: 487 mm), runoff and sorghum yields were:

Treatment (details in Table 61)	*Mean runoff* %	*Yields, kg/ha*	*Yield gain over control* %
1	18·0	1447	0
2	5·1	3215	122
3	6·1	2960	104

Further circumstantial evidence is provided by the report of Charreau and Poulain (1963; 1964) that at Bambey the storage capacity of plant-available water is the first limitation on millet production. On a dior soil (Ferruginous Soil on sand) with a storage capacity of only 125 mm rainfall, maximum yields were about 2 t/ha, whereas on a dek soil (Ferruginous-Vertisolic intergrade), storing 210 mm, 3·3 t could be achieved. Also in Senegal, Gillier (1964) showed the benefit of incorporating groundnut shells during land preparation on the subsequent groundnut crop, an effect limited to drier areas (Table 62) and attributed to better storage and utilization of water.

Another approach to the problem of moisture conservation in dry areas is suggested by the work of Schoch (1966). He calculated potential evapotranspiration for a number of sites on the Bambey farm and showed that the removal of trees (acacias at 25—30 per ha) from traditional farmland had allowed an increase in windspeed and hence in potential evapotranspiration. Differences between a bare field and a traditional field reached fifty per cent

TABLE 62

Effect of incorporating groundnut shells in the soil on subsequent groundnut yields (Gillier, 1964).

Site	*Rainfall (mm)**	*Year*	*Weight of groundnut shells incorporated, t/ha*						*LSD*
			0	5	7·5	10	15	20	(0·05)
Louga	450	1958	400	380				700	100
		1959	500	510				560	185
Tivouane	600	1959	1310	1770		1810		2100	340
		1960	1040	1150		1230		1470	230
Bambey	650	1960	3990		3980		3920		N.S.
		1961	2050		2430		2540		120
Darou	900	1960	2640	2620		2760			250
		1961	2300	2290		2580			N.S.
		1962	2130	2050		2270			400

*Rainfall data were not given in the original work, although the relationship between effectiveness of treatment and rainfall was pointed out. Figures given are approximate annual means and probably differ from those of the actual experimental years.

on some occasions. He therefore suggested the planting of windbreaks to produce a more favourable micro-climate for crop growth without hampering mechanization. However, the success of these proposals almost certainly depends upon the windbreak being of *Acacia albida*, a species which is not only leafless in the rainy season but which also appears to improve the fertility of the soil around it (see Chapter 6). The effect of its dry season transpiration on soil water balance appears not to have been calculated.

Water is also conserved in a bare fallow, and although evaporative loss occurs in the following dry season, the moisture deficit at the start of the next rainy season is less than after a crop (Kowal, 1969). However, in view of the labour needed to keep the fallow bare, the increased nutrient leaching losses that occur under bare soils (Tourte *et al.*, 1964) and, particularly, the high risk of accelerated erosion from an unprotected soil surface, the use of a bare fallow to conserve water cannot be justified.

In higher rainfall areas the choice is less clear, for despite a theoretical need to dispose of excess water, positive responses to moisture conservation techniques have been obtained even at sites with a humid period of more than two months. For instance, crop yield figures for the Niangoloko runoff plots suggest that the tying of contour ridges was beneficial under a mean annual rainfall of 1137 mm (Table 63). Outside the region, at Kongwa, Tanzania, on a "ferrallitic" soil, a system of tied ridges gave highest yields of maize not only in 1966-7 when rainfall was 548 mm but also in 1967-8 when it was 1090 mm (Macartney *et al.*, 1971).

The most extensive data in West Africa are those of Lawes (1957; 1961; 1963; 1966), working at Samaru. His hypothesis was that, because infiltration was reduced by soil capping, (a) on sloping ground, water was lost by runoff so that insufficient entered the soil, while (b) on level ground, ponding prevented adequate soil aeration, in both cases to the detriment of crop growth. The aim of his experiments was therefore to test the effect on crop yield of various measures to reduce runoff and improve infiltration. His results, summarized in Tables 64 and 65, show that:

1. Surface mulching improved the growth and yield of sunflowers very considerably and that of sorghum to a lesser extent. In the case of

TABLE 63

Runoff and crop yields from runoff plots at Niangoloko, Upper Volta, for the period 1956-60 (Fournier, 1967).

		Mean yields, kg/ha	
	Runoff, %	*Groundnuts*	*Millet*
Tied contour ridges	0·9	846	729
Open graded ridges	6·3	479	376
Flat cultivation	12·2	658	352

Site slope, 2·5%. Erosion data in Table 61.

TABLE 64

Effect of different methods of moisture conservation on crop yields at Samaru (Lawes, 1957, 1961, 1963).

1(a). Mulching of ridge-grown sunflowers with 8 cm depth of groundnut shells; 1956, rainfall 927 mm:—

	Yields	
	Green plants, t/ha	*Threshed seeds, kg/ha*
Mulched plots	15·2	349
Unmulched plots	6·8	95

1(b). Mulching of ridge-grown sorghum with groundnut shells or dead grass:—

		Grain yield, kg/ha			
Year	*Rainfall, mm*	*No mulch*	*Groundnut shells*	*Dead grass*	*S.E.* (±)
1957	1394	981	1250	1170	44
1958	889	1290	1689	1704	54

2. Cross-tieing to improve infiltration for ridge-grown cotton:—

		Yields, kg/ha		
Year	*Rainfall, mm*	*Open furrows*	*Tied ridges*	*Increase or decrease with cross-tieing, %*
1956	927	704	805	+14·3±2·6
1957	1394	965	818	—15·2±2·7
1957	1394	905	790	—12·7±2·7
1958	889	807	984	+22·0±2·7

sorghum the effect of mulch was significant in a wet year (1394 mm) as well as in a dry year (889 mm), a result suggesting that part of the improvement, perhaps a large part, was due to better aeration rather than to a greater conservation of moisture.

2. In relatively dry seasons (927 and 889 mm), ridge-tying increased cotton yields, presumably because it conserved water, as was found in East Africa (Dagg and Macartney, 1968); but in a wet season (1394 mm) it had a depressive effect.

3. The tying of alternate furrows only, or the tying of all the furrows in combination with a furrow mulching treatment, significantly improved

TABLE 65

Effect of measures to improve infiltration on crop yields at Samaru (Lawes, 1961, 1963, 1966).

Crop	Year	Rainfall, mm	Yields, kg/ha				S.E.(±)
			Open free-draining furrows	Alternate furrows cross-tied	All furrows cross-tied	All furrows cross-tied and mulched	
Cotton	1958a	889	891	1175	—	1160	53
	1958b	889	880	1034	—	1036	45
	1962	1303	1738	2136	1818	2190	76
Sorghum	1959	1062	2172	2329	2198	—	35
	1960a	1092	1988	2091	2001	—	60
	*1960b	1092	3237	3481	3374	3591	—
	*1962	1303	2398	2338	2549	2304	—
Groundnuts	1959	1062	1219	1261	1238	—	35
	1960	1092					
		Kernels	1624	1535	1536	1655	41
		Haulm	2306	2559	2637	2835	89

a, b—groundnut shell and grass mulches, respectively.
* figures are means of two different fertility levels.

yields of cotton over open furrows and over tied unmulched furrows in both wet and dry seasons (889 and 1303 mm). With groundnuts and sorghum the effects were much smaller and not significant. As Lawes (1966) pointed out, both these crops have a shorter growth period than cotton and, unlike cotton, both have passed maximum leaf area before the end of the rains.

The practice of tie-ridging under Samaru conditions was criticized by Kowal (1970b; 1907d) and Kowal and Omolokun (1970) on the grounds that from mid-season onwards rainfall greatly exceeds evapotranspiration. Even assuming a high initial soil deficit of 200 mm, in an average year there is an excess of about 250 mm rainfall occurring almost entirely in the two month period, August-September, which must either run off the surface or percolate through the profile (Kowal and Knabe, 1972). Once the profile is at field capacity, prevention of runoff can store water only by raising the water-table, which on lower slopes in a wet year is already close to the surface. Higher up the slopes there is the problem of slow infiltration and anaerobic conditions in the surface soil.

Lawes' results are not entirely at variance with Kowal's views. In a wet year (1957 with 1394 mm rain) tied ridges clearly depressed cotton yields, presumably because of waterlogging and anaerobic conditions in the rooting-zone, but in a dry year the response was positive. Moreover, the two compromise techniques, alternately tied ridges (to allow some runoff) and furrow mulching (to improve infiltration) conferred an advantage on cotton in both wet and dry years. One may conclude that in some years the bad effects of excess moisture during the humid period are either negligible or are outweighed by the benefit of a greater availability of moisture at some other time during the season.

For instance, Lawes (1966) showed that tie-ridging conserved moisture in the period before sowing. At 25 cm depth at Samaru on 20—21 May 1959, soil moisture contents, as a percentage of oven-dry soil weight, were:

	Expt. 1	*Expt.* 2
All furrows open	12·5	12·0
All ridges tied	17·7	19·1

This was in a soil in which the wilting point and field capacity are 12 and 19 per cent, respectively. Rainfall at this period is unreliable, and increased soil storage through prevention of runoff can benefit early-planted crops (cereals, groundnuts) in seasons in which sowing is followed by a drought.

However, at Samaru the most responsive crop to moisture conservation practices was cotton, which is not planted until the rains are well established. For this crop any benefit must come from better moisture conditions at the end of the season. As an indeterminate crop, cotton continues to grow and flower as long as there is sufficient moisture, so that any practice which helps to conserve the last few rains of the season may be expected to increase yield. In fact, the increase is not always appreciable, because in some years the rains end abruptly at a time when the soil profile is at or near field capacity.

(d) *Conclusions*

In the management of soil water two main aims must be reconciled (i) to keep soil moisture optimal for crop growth, and (ii) to prevent loss of soil by erosion. Central to both aims is the control of runoff. The maintenance of optimal soil moisture has priority, but once this requirement has been met any excess water should be disposed of in the least erosive way.

In the driest part of the region both aims are realized by preventing any runoff, and hence holding all rain where it falls. In the wettest areas it is probably always necessary to permit some runoff. In between, it seems, there is an area in which measures to conserve water are useful only at certain times and for certain crops. Any clear delimitation of these areas must, however, await the collection of a great deal more agroclimatological and crop physiological information than is at present available.

From the existing, rather limited, experimental data, it appears that even where mean rainfall exceeds 1000 mm in a five-month season, measures which reduce or prevent runoff can give some advantage (a) to early-planted crops when the rains start slowly and erratically, and (b) at the end of the season, to a crop like cotton with a growth period extending well into the dry season; but these advantages are highly dependent upon the pattern of the rainfall and will not occur every year. Whether management techniques should be adopted that prevent runoff over the whole season depends on circumstances. These may be given as (a) the crop to be grown, (b) the water relations of the soil, (c) the position on the catena and, hence, the depth to the wet-season water-table, and (d) the local distribution of rainfall, which is probably the most important factor of all. As the latter may show a great annual variation, management should above all be flexible, so that where feasible, ridge-ties can be taken out promptly if they are causing serious waterlogging and put in promptly when the rains appear to be ending. In these circumstances the cost and availability of labour may prove to be the overriding factor. A further possibility at some sites is the use of supplementary irrigation at the end of the season.

Alternatively, one may grow crop varieties of sufficiently short growth cycle to reduce the need for elaborate measures to conserve moisture. Soil water management is then mainly a problem of erosion control.

It is also possible, as Lawes (1961, 1963, 1966) showed, to prevent runoff and avoid serious waterlogging even in years of high rainfall by a vertical mulching technique. Slits cut in furrow bottoms and filled with groundnut shells gave greatly improved infiltration. However, this technique requires mechanization. A second disadvantage is that leaching may be increased, and though this may have no immediate effect on crop yield, it will cause a gradual depletion in soil nutrient reserves.

In southern areas the pattern of water availability is complicated by the two-peak nature of the rainfall. In some areas there is a period of up to a month in the middle of the rainy season when mean rainfall barely equals potential evapotranspiration (Kowal and Knabe, 1972). In drier years long-season crops may suffer a period of water-stress at this time. This can be avoided by growing two short-season crops, though the yield of the second crop is often low due to the rains ending before maturity. Unfortunately, no reports of moisture conservation studies in these areas have been found.

Where there is surplus water to dispose of, the most important principle is to keep its rate of flow to a minimum. For large-scale mechanized operations graded terraces with a well-protected system of grassed watercourses are essential for erosion control. If the terraces have been well levelled and are not too wide, flat cultivation may give good soil protection; otherwise, ridges set slightly off the contour may be more satisfactory. Alternatively, one may augment the effect of the banks between terraces with narrow grass buffer strips at intervals between the banks. This permits wider terraces to be used with safety.

On small farmer's fields ridges slightly off the contour are likely to give the greatest protection, in conjunction with grass buffer strips between fields to prevent large-scale movements of soil. Where mounds are preferred, mulching and tying will control erosion, though the ponding that this may involve could be harmful in wetter areas.

2 CULTIVATIONS

Satisfactory plant growth depends upon an effective rooting system. Any physical factor tending to limit root extension is liable to reduce uptake of moisture and nutrients and so limit crop growth and yield. It follows that the maintenance of soil fertility is concerned as much with keeping the soil in a physical state that favours effective rooting as it is with maintaining nutrient status.

There have been many reports from the savanna of the deterioration in soil structure that accompanies cropping. As Charreau and Nicou (1971a) put it " . . . . no natural factor is favourable to the development of structure. Rather the contrary: the exceptional aggressiveness of the rains, the dryness and length of the dry season, the sandy or clayey-sand texture of the surface horizons and the predominance of kaolinite among the clay minerals are some of the factors encouraging the degradation of the cultivated profile and the disappearance of structure". Thus in utilizing these soils, particularly if permanent agriculture is envisaged, it is essential to employ management practices which allow for the harsh environment and which are designed to create or improve soil structure.

In some of the francophone countries of West Africa research on cultivations is already well advanced. This is partly because in Senegal in particular there is only a short period between the onset of the rains and the optimum planting date (Chapter 1). Much of the existing literature has been reviewed by Charreau and Nicou (1971a, b, c, d). The following account is based in large part on their findings.

(a) *Soil physical properties and crop rooting*

The soil physical parameters most closely linked with the development of crop roots are the interrelated properties of porosity and bulk density. Various hypotheses have been put forward to account for the beneficial effects that increases in porosity often have on rooting. First, there may be differences in the soil moisture regime; permeability is better at lower bulk densities, so that the early rains may penetrate more easily, and changes in porosity and in pore-size distribution may increase available water; secondly, aeration is generally better; and thirdly, and probably most important in the present context, a decrease in bulk density reduces mechanical resistance to root penetration. Work elsewhere in this field has been reviewed by Barley and Graecen (1967).

Bulk densities of savanna soils are often high. Charreau and Nicou (1971a) reported values in the range 1·60—1·70 in unploughed surface soils in Senegal; and at Samaru, just below the plough-layer, at 15-23 cm, bulk densities are about 1·53 (41·2 % porosity) in material classified as a sandy clay or sandy clay loam (Kowal, 1968a). In both cases a risk of restricted rooting seems possible. However, although 40 per cent porosity has often been cited as the threshold value below which rooting is impeded, Veihmeyer and Hendrikson (1948) showed that it depends very much on the type and texture of the soil. In an experiment with sunflowers the threshold bulk densities above which roots failed to penetrate were about 1·75 for sandy soils and ranged between 1·46 and 1·63 for clays. Bulk density is therefore only a very approximate measure of the restriction on root growth. More important is the distribution of pore sizes. For good growth continuous pores of about 0·2 mm are needed; but few, if any, studies of cultivations in the savanna have included measurements of pore size distribution.

Evidence that soil compaction does limit rooting depth has come from the work of Bertrand (1971) at Bouake, Ivory Coast. He showed that the loosening of deep horizons encouraged a much greater root extension of upland rice to below 60 cm; and it seems probable that the decline in the rooting density of rice with the length of time the soil had been under hand cultivation, noted by Siband (1972), was also related to the high bulk densities of long-cultivated fields.

Quantitative studies in Senegal showed a very rapid decline in root weight and rooting densities with increasing bulk density (Charreau and Nicou, 1971b). For example, regression equations for the rooting density of maize in the 10-20 cm horizon on soil bulk density measured at or shortly after planting were:

at Sinthiou-Maleme: $y = -1 \cdot 96x + 3 \cdot 46$ ($r = 0 \cdot 89^{***}$)
and at Sefa: $y = -4 \cdot 85x + 7 \cdot 71$ ($r = 0 \cdot 988^{***}$)
and for sorghum at
Nioro-du-Rip: $y = -2 \cdot 45x + 4 \cdot 15$ ($r = 0 \cdot 965^{***}$)

where y = rooting density in g/dm^3, and x = bulk density in g/cm^3. At Bambey too, Blondel (1965a) obtained significant regressions of both root weight and crop yield on soil bulk density that were linear for sorghum and hyperbolic for groundnuts (Fig 9). Decreases of sorghum yield amounted to 450—600 kg/ha (or 14-20 per cent) per 0·10 unit increase in bulk density. For groundnuts loss of yield between bulk densities 1·50 and 1·60 was 730 kg/ha (about 55 per cent.)

The dependence of yield on bulk density follows from the relationships between rooting and bulk density and between top growth and root growth. Other studies in Senegal have shown the importance of this latter relationship. Linear correlations between root weights and crop yields were found at three different sites, Nioro, Bambey and Sefa, and for three different crops, maize, groundnuts and sorghum (Charreau and Nicou, 1971b). For example, at Nioro the relationship was obtained:

$$\text{Sorghum yield (kg/ha)} = 1110\,R + 3122 \quad (r = 0 \cdot 713)$$

where R = rooting density in 10-20 cm horizon in g/dm^3. The correlation was even higher when expressed in terms of the percentage (P) of the root system found in the 10-20 cm horizon (Nicou and Thirouin, 1968):

$$\text{Sorghum yield (kg/ha)} = 55P + 3040 \quad (r = 0 \cdot 831)$$

(b) *Cultivations and the physical properties of the rooting zone*

Crops root better in a loosened soil, and, as will be seen, there is often a yield response to ploughing. However, the relationship between soil cultivations and crop growth is not a simple one, and bulk density is not the only factor involved. This section examines the effect of cultivation on the physical conditions in the rooting medium and how they can affect root growth and root distribution in the profile.

The effect of ploughing on soil bulk density is well documented for Senegalese conditions. Table 66 gives data from four sites. At all of them ploughing lowered bulk density and increased porosity from 38—43 per cent to 42—48 per cent. But these measurements were made soon after the first rains, following ploughing at the end of the previous season. Persistence of the increased porosity produced by ploughing varies widely. It may be for the whole season (Nicou, 1969) or for only a few weeks, depending on soil, crop cover and the violence of the rainfall. In one groundnut experiment at Bambey the bulk density of ploughed soil had already reverted

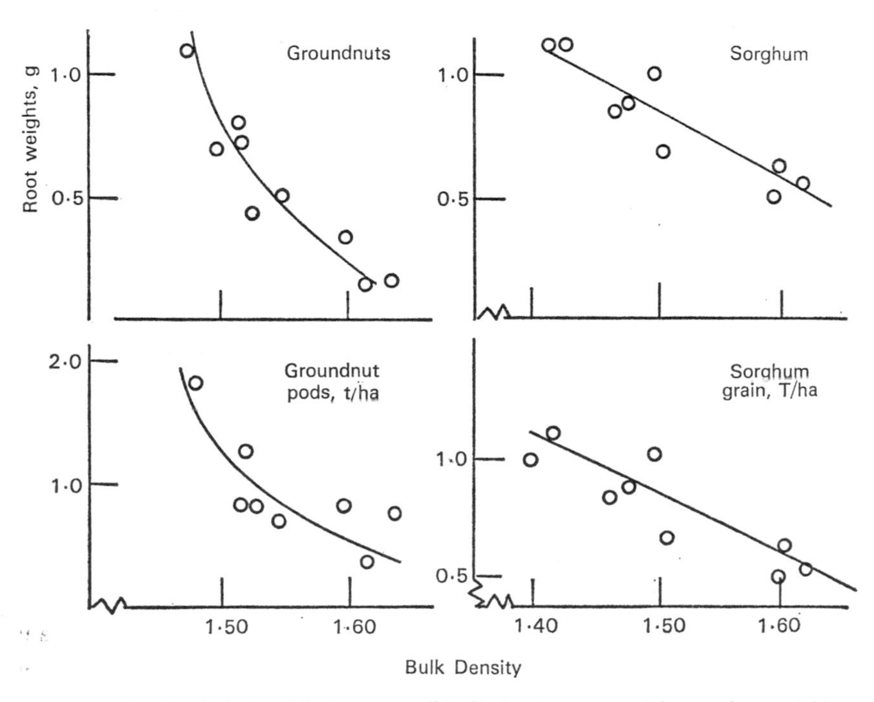

Fig. 9. Relationship between soil bulk density, root weights, and crop yields at Bambey, Senegal (Blondel, 1965a).

TABLE 66

Comparison of topsoil bulk densities of ploughed and unploughed fields at four sites in Senegal (Charreau and Nicou, 1971c).

Site and soil	*Year*	*Bulk density, g/cc*		*Level of statistical signif.*
		Unploughed	*Ploughed*	
Bambey, dior, (v. sandy Ferrug.)	1964	1·62	1·50	0·01
	1965	1·64	1·47	0·01
	1967	1·57	1·51	0·01
	1968	1·54	1·54	N.S.
Bambey, dek, (clayey sand)	1964	1·61	1·43	0·01
Sinthiou-Maleme (Ferrug.)	1967	1·52	1·50	N.S.
	1968	1·53	1·49	N.S.
	1969	1·49	1·48	N.S.
Nioro-du-Rip, (Ferrug.)	1967	1·50	1·40	0·01
	1968	1·52	1·43	N.S.
	1969	1·54	1·42	N.S.

from an initial value of 1·47 to 1·64 (equal to the unploughed control plot value) one and a half months after the start of the rains, when the cumulative rainfall was only 180 mm (Charreau and Nicou, 1971c). However, even a short-lived improvement in porosity early in the growth cycle may be sufficient to make a significant difference in the eventual crop rooting pattern.

Ploughing also changes the soil moisture regime. Charreau and Nicou (1971c) discussed the mechanisms behind the common observation that crops on ploughed soils generally resist dry periods very much better than the same crops on unploughed soils (except when the drought is very severe). First, and this applies only where ploughing is done at the end of the rains, the creation of a superficial mulch of soil crumbs by a well-timed cultivation may reduce evaporation losses and conserve moisture during the dry season. Figures from Samaru showed that ploughing at the end of the rains reduced moisture loss from the top 75 cm of soil by 38 mm, though this appeared to be due very largely to the curtailment of weed growth (Jones, 1974b).

Secondly, ploughing usually improves infiltration. For example, runoff measurements at Sefa in 1968 (Charreau and Seguy, 1969) showed that 578 mm of the 729 mm rainfall infiltrated into a ploughed soil, but only 459 mm into an unploughed soil. However, as Charreau and Nicou (1971c) pointed out, the difference is likely to be much smaller in areas of gentler relief and lower rainfall, where substantial runoff is rare, as at Bambey.

Thirdly, the utilization of soil moisture is closely related to root distribution, and changes in the pattern of root distribution due to changes in bulk density inevitably affect the soil moisture profile. This was illustrated by a study comparing the effect of mouldboard ploughing (30 cm) and shallow surface tillage (control) on the rooting, top growth and yield of four varieties of upland rice in Casamance, Senegal (Nicou *et al.* 1970; Charreau and Nicou, 1971c; Haddad and Seguy, 1972). Ploughing approximately doubled the number and length of the main roots relative to the control, increased their degree of branching and the proportion of secondary roots, and increased the estimated root surface area by between 50 and 150 per cent. This increase occurred at all depths, but was particularly large in the 10—20 cm horizon; one variety showed a twelve-fold increase in this horizon relative to control.

The effect was seen in the first year of the experiment, when the crop was subject to a prolonged drought at the stage of rapid vegetative growth. Plants on ploughed soil showed greater drought resistance, and examination of the moisture profile at the end of the drought period showed why. The moisture content of the ploughed profile had been reduced to a much lower level than that of the control profile, the difference amounting to a rainfall equivalent of 30—40 mm. Whereas on ploughed plots available water had been exploited down to 100 cm, in unploughed plots exploitation was limited to the top 20 cm. Even in the second year of the experiment, when there was no comparable drought stress, clear evidence was obtained of a much deeper use of soil moisture in ploughed soils.

In this case and in many others, ploughing to 30 cm encouraged crop roots to penetrate below 30 cm, but some exceptions have been reported. At Sefa, Deffontaines (1965) observed that ploughing often created a discontinuity in the profile which could act as a barrier to root development; and at the same station, Blondel (1966) observed that the roots of a millet crop were concentrated in the 0-20 cm horizon after ploughing. However, Charreau and Nicou (1971c) regarded the limitation of rooting by "plough pans" as rare and likely to occur only in soils with a high concentration of clay at depth.

Data from many other root distribution studies on other crops and at other sites in Senegal, collected by Charreau and Nicou (1971c), show a frequent rooting response to ploughing. Although this response is often large (Table 67), it does not always occur. Rooting may even be reduced where cultivation has loosened the soil too much just before sowing (Nicou and Thirouin, 1968). Careful judgement is required to obtain the maximum effect from soil cultivation.

(c) *Effect of method of cultivation on yields*

Since crop yields depend upon root growth, and root growth may be increased by ploughing, a strong relationship between crop yield and the type of soil cultivation may be expected. Such a relationship has often been found.

Summarized results from experiments in Senegal, Upper Volta, Niger and Mauritania are given in Table 68(a). They show a positive (though not necessarily always significant) response to ploughing in 100 out of 112 experiments, involving five different crops, with a mean yield increase of about

25 per cent. In these experiments the comparison was between ploughing (usually mouldboard, depth 15—25 cm) and surface tillage or hand hoeing. Other work has compared the effects of a range of cultivation depths. Results from Saria, Upper Volta, for example, showed a gradual increase in sorghum yield and dry matter production with increasing depth of cultivation (Table 69).

TABLE 67

Comparison of the effects of ploughing and superficial tillage on crop rooting at Bambey, Senegal (Blondel, 1965b).

Soil	*Crop*	*Sampling depth, cm.*	*Comparative root weights*	
			Superficial tillage	*Ploughing*
Dior (v. sandy)	Groundnuts	0—10	35	138
		10—20	32	148
Dek (clayey sand)	Sorghum	0—10	123	201
		10—20	56	97
		20—40	16	23

TABLE 68

Effect of ploughing, with and without the incorporation of plant material, on crop yields (Charreau and Nicou, 1971c).

Crop	*Number of expts.*		*Mean yields, kg/ha*		*Increase, %*
	Total	*Positive*	*Control (shallow cult'n)*	*Increase from ploughing*	
(a) Ordinary ploughing (no incorporation):—					
Millet	22	21	1245	254	21
Sorghum	46	39	1874	536	29
Maize	6	6	2093	568	27
Cotton	7	7	1629	433	27
Groundnuts	31	27	1412	274	19
(b) Ploughing with incorporation:—					
Millet	5	4	971	365	38
Sorghum	2	2	2039	532	26
Maize	12	10	1474	970	66
Cotton	12	10	1240	423	34
Groundnuts	113	81	1661	119	7

n.b. Since sites and experimental years were often different no direct comparisons can be made between the two halves of the table, a and b.

TABLE 69

Effect of cultivation depth on the growth and yield of sorghum on a leached Ferruginous Soil at Sarla, Upper Volta (IRAT/Haute Volta, 1972).

Cultivation treatment	*Cultivation depth, cm.*	*Yields, kg/ha*	
		Grain	*Stover*
Tractor ploughing			
— end of rains	21	2319	8786
— start of rains	21	1491	9784
Ox ploughing	15	2364	8243
Donkey ploughing	9	2129	7953
Surface harrowing	<5	2130	6286
Hoeing	<5	1764	6014

TABLE 70

Effect of ploughing depth on millet and sorghum yields and response to nitrogen at Bambey, Senegal (Poulain and Tourte, 1970).

		Crop yields, kg/ha			
		1966, *Millet*		1967, *Sorghum*	
Ploughing treatment	*Fertilizer treatment***	*Dior soil*	*Dek soil*	*Dior soil*	*Dek soil*
Shallow surface tillage	Control	684	935	320	999
	N_0	890	1363	349	1607
	N_1	1143	1658	427	2155
	N_2	1253	1698	426	2620
	N_3	1203	2026	410	3071
Ploughing*	Control	840	1146	390	2042
	N_0	881	1402	543	2090
	N_1	1305	2248	602	2481
	N_2	1516	2183	660	2752
	N_3	1622	2051	1078	2949
Deep ploughing*	Control	1069	863	553	1851
	N_0	1063	1013	956	2274
	N_1	1468	1817	1202	2698
	N_2	1618	2140	1585	2944
	N_3	1391	1744	1753	2843

*Ploughing involved subsoiling to 40 cm (dior) or 30 cm (dek), then mouldboard ploughing to 25-30 cm (dior) or 20 cm (dek), followed by rotary hoeing. Deep ploughing involved subsoiling to 70 cm (dior) or 60 cm (dek), then mouldboard ploughing to 35-40 cm (dior) or 25-35 cm (dek), followed by rotary hoe (dior) or crosskill harrow (dek). Treatments were the same in both years so that 1967 results may have been the consequence of cumulative effects.
**Control—no fertilizer at all. All other treatments received PKS and trace elements. N-levels were 0, 50, 100, 150 kg N/ha.

Rather more spectacular were the results of Poulain and Tourte (1970), who investigated the effect of very deep ploughing (60-70 cm) on the yields of millet and sorghum and their response to nitrogen at Bambey (Table 70). They found that

(i) in the absence of any fertilizers, the response to ploughing was small to moderate for both millet and sorghum on the dior soil (very sandy), but large for sorghum on the heavier dek soil;

(ii) there were moderate to large responses to nitrogen irrespective of depth of cultivation, but with the important exception that

(iii) sorghum responded to nitrogen on the dior soil only after ploughing, the response with deep ploughing being markedly greater than that with shallow ploughing.

No interpretation was offered for this latter interaction, but in the light of the later study of Blondel (1971c), which revealed very rapid rates of nitrate leaching in dior soils, it seems probable that deep ploughing was effective because it enabled the sorghum roots to go deeper and recover nitrate washed into the subsoil.

Although much of the data presented so far has come from the drier areas of Senegal and, in particular, from Bambey, the influence of ploughing on yield is not limited to semi-arid areas with very sandy soils. Of the results already quoted, the six maize experiments in Table 68a were all sited in Casamance on either Ferrallitic Soils or leached Ferruginous Soils with heavy subsoils, receiving about 1300 mm annual rainfall; the data in Table 69 came from a leached Ferruginous Soil with 850 mm per annum; and the ploughing which improved rooting density and drought resistance in the Sefa rice experiment described above also significantly increased yields (Table 71).

Results from other West African cultivation experiments (Table 72) show positive responses to ploughing in the forest zone no less than in the dry savanna. Le Buanec (1972) noted that in southern and central Ivory Coast the effectiveness of ploughing depended on the conditions under

TABLE 71

Effect of two seasons of soil cultivation on the yield of upland rice at Sefa, Senegal (Haddad and Seguy, 1972).

		Yields, q/ha, after the following treatments:—			
Soil type	Year 1: Year 2:	*Control* *Control*	*Control* *Plough*	*Plough* *Control*	*Plough* *Plough*
Slightly Ferrallitic		15·7	23·7	23·5	29·0
Leached Ferruginous		0·0	27·0	30·2	32·3
Hydromorphic		0·0	18·9	18·5	23·5

Control treatment was shallow surface tillage.

TABLE 72

A summary of yield responses to ploughing at seven sites in West Africa.

Site and mean annual rainfall, mm	*Soil description*	*Crop*	*Number of expts.*	*Treatments*	*Mean yields kg/ha*	*Increase over control, %*
1. Kwadaso, Ghana (1500)	Well drained sandy loam	Maize	4	Hoe	1374	—
				Plough (23 cm)	1580	15
				Plough (38 cm)	1538	12
		Cassava	1	Hoe	18875	—
				Plough (23 cm)	24795	31
				Plough (30 cm)	24857	32
2. Man, Ivory Coast (1715)	Ferrallitic	Rice	1	Hoe	775	—
				Plough	1160	50
		Maize	1	Hoe	590	—
				Plough	810	37
3. Ferkesse-dougou, Ivory Coast (1715)	Ferrallitic	Rice	2	Hoe	1243	—
				Plough	1110	11
		Maize	2	Hoe	1758	—
				Plough	1795	2
4. Bouake, Ivory Coast (1150)	Ferrallitic	Rice	3	Hoe	967	—
				Plough	1117	16
		Maize	3	Hoe	1558	—
				Plough	1887	21
5. Farako Ba, Upper Volta (1100)	Slightly Ferrallitic	Cotton	2	Hoe	933	—
				Harrow	990	6
				Plough	1155	24
		G'nuts	1	Hoe	2461	—
				Harrow	2732	11
				Plough	2811	14
		Sorghum	1	Hoe	1077	—
				Harrow	1220	13
				Plough	1517	41
6. Saria, Upper Volta (900)	Ferruginous	Sorghum	3	Hoe	1393	—
				Harrow	1711	23
				Plough	1931	39
		Cotton	1	Hoe	1593	—
				Harrow	1586	0
				Plough	1652	4
7. Guetele, N. Cameroun (820)	Weakly developed Dune soil	Sorghum	2	Plough v. hoe	—	30
		Sorghum	2	Plough v. hoe	—	76

Sources: 1. Ofori and Nandy, 1969; Ofori, 1973. 2, 3, 4. Le Buanec, 1971; 5, 6. IRAT/Haute Volta, 1972. 7. Vaille, 1970.

which it was carried out. If they were good, then a good depth of loose porous soil was produced, which, except where drought intervened, improved yields. But if the land was ploughed when too dry, hard clods of low porosity were formed, and root distribution and top growth of the subsequent crop were poorer than in the unploughed control (see results of Ferkessedougou). This may be less of a problem in the sandier, less cohesive soils of the more arid areas. Tourte *et al.* (1963) recommended dry ploughing for millet cultivation in sandy soils receiving less than 800 mm per annum, but not for sorghum growing in heavier soils.

(d) *Residual effects from ploughing*

Although improvements in porosity from ploughing are only temporary and bulk densities, particularly of very sandy soils, may soon revert to their original high values, in some soils a residual effect, sufficient to influence rooting and crop yields, persists into the second season (Table 73). For instance, the yield data in Table 71 imply a very considerable residual, as well as cumulative, effect from ploughing in the second-year rice crop at Sefa.

Charreau and Nicou (1971c) discussed these residual effects in some detail and concluded that the main factors, apart from soil texture, that affect the preservation of structural improvements into the second season are:

(i) The volume and intensity of the rainfall in the first season, through its direct effect on soil compaction and indirectly through its effect on crop growth.

(ii) The form and condition of the initial cultivation: low bulk density is more likely to survive if the cultivation produced a cloddy rather than a fine

TABLE 73

Effect of ploughing on soil porosity at the beginning of the second season after ploughing (Charreau and Nicou, 1971c)

Site and Soil	*First season's crop*	*Porosity, %, in 5—20 cm horizon*		*Significance level of difference*
		Control	*Plough*	
1. Bambey				
-dior	Sorghum	39·2	41·5	0·05
-dek	Sorghum	38·1	40·4	0·05
2. Sefa				
-sl. Ferrallitic	Rice	41·0	50·0	0·01
-Ferruginous	Rice	40·0	50·0	0·01
-sl. Ferrallitic	Rice	44·0	52·0	0·01
-Hydromorphic	Rice	46·0	51·0	0·01
3. Bambey				
-dior	Groundnuts	39·6	38·9	N.S.

tilth. Further, where the cultivation involved the incorporation of plant residues the improved porosity is generally better preserved (see below).

(iii) Cropping practice: time of planting is particularly important because it affects the degree of protection the soil receives from rainfall impact, but other operations like weeding are also involved. For instance, the harvesting of groundnuts tends to destroy soil structure.

(iv) Nature of the crop is important because it affects the degree of soil cover. Rooting habit is also relevant; roots of cereals, except maize, appear to preserve structure better than those of cotton or groundnuts.

(e) *Cultivations involving incorporation of plant material*

Several references have already been made to the practice of ploughing in plant material. Depending on the circumstances, the material may be either crop residues, the standing vegetation of an herbaceous fallow or a specially grown green manure crop. The following observations derive largely from reviews by Charreau and Fauck (1970) and Charreau and Nicou (1971c, d) of experimental work on this practice carried out in Senegal. Very little appears to have been done elsewhere.

Because ploughing usually improves crop growth and yield, ploughing at least once per rotation is regarded as an essential feature of any improved agricultural system. Where the ploughing operation also involves the incorporation of plant material, yield increases, except of groundnuts, are similar or greater than those obtained from ordinary ploughing (Table 68b), and there is strong evidence that the structural improvement with incorporation of plant residues is more long-lasting. Charreau and Nicou (1971c) described the organic material as forming initially a framework within the otherwise poorly structured soil. Breakdown of this organic matter by the mesofauna (largely termites) creates a "breadcrumb" structure within the soil, and the original framework persists into the following season as a network of empty spaces that can be used by roots.

Incorporation of crop residues is only possible when a short-season crop is grown. Legume haulm is removed for fodder, and long-season cereals mature after the end of the rains, when the soil is too dry and hard to cultivate. Incorporation at the beginning of the next season is not regarded as practicable because the residues would not be sufficiently well decomposed at sowing time. Shortening of the growing season of traditional cereals to permit residue incorporation has therefore become one of the aims of the breeding programme in Senegal. One may add that, in a mechanized system, incorporation is probably the easiest way of disposing of unwanted residues, quite apart from any advantage they may confer on the soil.

Green manuring is controversial; some of the arguments against it were given in Chapter 10. But it may have a useful function in areas where a short rainy season is preceded by a short preparatory period, as is the case in much of Senegal. Ploughing at the beginning of the rains delays planting and reduces yields, but ploughing at the end of the rains is possible only after a short-season crop. Under these conditions, if land is ever to be cultivated deeply, there must be set aside one year per rotation in which the farmer can plough at a time of his own choosing. However, the land must not be left bare because of the erosion hazard, and the choice of covering vegetation lies between a volunteer fallow of herbaceous species and a green manure crop. The ploughing in of this vegetation may be achieved in one operation using an inverting plough. The timing of this operation is important; it should not be too early, because that would leave the soil exposed to the late rains, but nor should it be left until the soil is drying out. It has been found that, providing the soil is near field capacity at the time of ploughing and the plant material is not too lignified, decomposition proceeds well even if no more rain falls. In addition, the simultaneous incorporation of a high rate of rock phosphate with the fallow or green manure appears to increase the effectiveness of the phosphate.

No systematic difference between the effects of green manure and of incorporated fallow have been reported; but whereas the fallow is usually allowed to occupy the land for two or three years, the green manure crop is always for one season only and so keeps the land out of production for a shorter period. Against that must be set the fact that the green manure crop requires land preparation, sowing and usually, if it is to be effective, some mineral fertilization. Hence, this system has little to recommend it to the ordinary farmer. A more promising alternative is to use the fallow year to grow a fodder crop, which may be cut about one month before the end of the rains, and then the regrowth may be ploughed in as green manure (Roose, 1967; Charreau and Nicou, 1971d).

(f) *Conclusions*

There have been many reports of increased crop yields as the result of ploughing. The main reason appears to be that loosening the soil encourages root extension. Such a relationship is not inevitable, for if a crop can obtain all the water and nutrients it needs from a small rooting volume, longer roots will give no benefit. However, few savanna soils provide for the needs of a high-yielding crop within a small rooting volume; nutrient levels are generally low and leaching is rapid in sandy soils (and most topsoils are sandy). Even where nutrient deficiences are overcome by mineral fertilization, water supply is often limiting, the more so because fertilizers usually increase crop leaf area; the moisture holding capacity of most soils is low, and rainfall is often erratic. For these reasons, a deep and extensive root system is usually an advantage.

Although there are wide variations with soil type, the loosening effect of a soil cultivation is short-lived. Under heavy rain a sandy soil may pack down to its original bulk density within a month or two, particularly if it is not well covered with a crop. However, there is now a considerable body of evidence from Senegal indicating that the duration of the ploughing effect may be increased if the cultivation includes the ploughing in of plant material, but the operation must be well timed and well executed.

All these findings beg the question of tractive power. Best results have been obtained under mechanization, and this is essential for incorporating large quantities of organic residues. However, as Table 69 shows, ox ploughing, though shallower than that achieved with a tractor, can increase yields and constitutes a worthwhile advance over traditional superficial hoeing.

Chapter 12

PRESENT KNOWLEDGE AND FUTURE PROSPECTS

Our subject is the maintenance and improvement of the fertility of savanna soils, but actual productivity, and hence the demand made by the crop on the soil fertility, is a function of many other factors besides soil fertility itself. The level of fertility to be maintained therefore depends on the system of land-use and the standard of husbandry to be employed.

In the past, and to a large extent to the present day, the savanna cultivator has been able to support himself because his farming system makes only light demands on the soil nutrient resources. His crop varieties are adapted to low fertility, planting densities are low, ashes and organic residues are generally returned to the fields and, above all, most of the land is allowed to revert to bush fallow for a long period after only a few years of cropping. This system maintains fertility at a low, but adequate, level. Already, however, population and economic pressures have begun to change the pattern in some areas; fallows have shortened or disappeared, and the introduction of cash crops has encouraged the exploitation of soil resources.

These social and economic changes will not be reversed, and more likely they will accelerate and spread. The task of the agronomist and soil scientist in this situation is twofold: first, to assist the gradual development of the existing agricultural system (through the supply of new crop varieties, pest control measures and fertilizer recommendations, preferably as one package deal) to enable the farmer to meet the immediate and increasing demands of the community; and secondly, to devise viable, high-productivity systems of agriculture, which utilize the land on a stable, permanent basis, and towards which the existing system may be expected to evolve as more capital becomes available for investment in agriculture. As Tourte (1971) has emphasized, these two types of development should not be regarded as opposed or competitive, but rather complementary, both chronologically and geographically.

In this chapter we first summarize the findings of earlier chapters and then relate them to agricultural development (a) in comparatively low intensity farming where traditional methods of maintaining fertility, e.g. bush fallowing, are still practised and (b) in more intensive farming systems, in which yields need to be high to justify a higher capital outlay.

1 SUMMARY

(a) *Climate*

In many respects the agricultural environment is a harsh one. High temperatures and high solar radiation encourage rapid growth, but crop production is often limited by insufficient water, for almost everywhere annual potential evapotranspiration exceeds rainfall. While complete crop failure due to prolonged drought is rare except in the marginal areas of the north, over a much wider area the shortness of the season limits both the range of crops that can be grown and their productivity, and yield reductions from short periods of drought during the growing season are common.

Over much of the savanna the rainfall is concentrated into a rainy season of between 3 and 6 months, and even in drier areas there are periods during which precipitation exceeds evapotranspiration. Hence, the agriculturalist may be faced with the need for water conservation measures at the beginning and end of the season, but with problems of disposal of surplus water in the middle. Moreover, the rainfall is often violent. Rates in excess of 100 mm per hour are common, and the erosion hazard is in consequence high.

(b) *Soil physical properties*

Natural soil fertility tends to be low. The land surface is old, and much of the material of present-day soils has been subject to more than one cycle of pedogenesis. Base-rich primary rocks are rare, and soil parent materials are mainly acid crystalline rocks or sedimentary rocks, alluvium, colluvium and aeolian deposits derived from them. As a result, most soils are sandy and low in crop nutrients, and the texture of their topsoils has been made coarser by the widespread phenomenon of downward eluviation of clay within the profile. Except in the more arid areas and in Vertisols the clay is predominantly kaolinite.

Because of their sandy texture many topsoils have poor moisture holding capacity. The effect of this on crop growth may be aggravated by restricted rooting depth, due to shallowness of profile or, in many Ferruginous Soils, to the presence of a concretionary, indurated, or compact clay horizon.

Soil structure is generally weak. Clods produced by cultivation show little resistance to rain action, and once these have been broken down the surface dries as a crust, restricting infiltration, aeration and seed emergence, and increasing runoff. Two cohesive factors are involved in aggregate formation, organic matter and free iron oxides. Of the former the unhumified fraction appears to be the most important, and there is evidence that an improved structure can be maintained by frequent incorporation of organic materials. Free iron oxides may bind soil particles into relatively stable crumbs, but those Ferrallitic Soils in which this mechanism appears to be important occur only in the most southerly parts of the region.

Any structure that exists in the topsoil under natural vegetation rapidly deteriorates under cultivation. Soils become single-grained and close-packed under the action of the rain, and many long-cultivated fields have high soil bulk density and low porosity, which are thought to restrict rooting.

Except after recent tillage, infiltration rates of cultivated soils are low, and during the humid period, when rainfall exceeds evapotranspiration, subsoil horizons of low permeability may cause surface waterlogging of some Ferruginous Soils. Hence, in many areas runoff is almost inevitable during the heaviest storms.

The extent of erosion in the savanna is poorly documented. Gullying, which is a serious problem in some areas, often follows the clearing of the natural vegetation; but erosion caused by raindrop splash and sheet flow may be more general. It is therefore of great importance to establish a complete crop cover as early in the season as possible, and for this an adequate level of fertility is essential. Even the mixed crops of the traditional farmer often fail to cover the soil adequately because poor fertility compels a low planting density. Material loosened by raindrop impact is carried away by the runoff water, but the scour of the runoff water itself is another cause of erosion, and a low velocity of runoff is essential to soil conservation.

Runoff and erosion involve the loss of soil and soil nutrients. Data are available from only a few sites, but it seems clear that the material lost is usually richer in nutrients than that left behind. There is another aspect: in some parts of the world a certain amount of soil loss through erosion is acceptable, because soils are deep and profiles fairly uniform in their physical and chemical characteristics; but in the savanna many soils overlie relatively soft, sesquioxide-rich horizons, which may harden irreversibly on exposure to dry air, and any soil loss brings nearer the time when the land will become unproductive. At other sites soils overlie already hard concretionary horizons, and any erosion here reduces rooting depth.

(c) *Soil chemical properties*

With the exception of some Vertisols and Eutrophic Brown Soils, soil chemical status is poor or very poor. For geological and pedological reasons soil total phosphate and topsoil clay contents are generally low and, except in more arid areas, kaolinite dominates the clay fraction. Because of the climate and the low clay contents, soil organic matter contents are also low; and, as a result, most soils have a very limited cation exchange capacity.

In consequence, phosphate deficiency is widespread, amounts of organic nitrogen, phosphorus and sulphur mineralized annually are often well below the requirement for good crop yields, and the soil is poorly buffered against changes induced in it by agricultural activities (e.g. crop nutrient removal, increased leaching, return of ashes and application of fertilizers).

Agronomic studies have demonstrated acute phosphate deficiency in most parts of the savanna. In some areas, in the absence of fertilizer, crop production is possible only as a result of the release of available phosphate by bush burning. As yet, no widely applicable method exists for the estimation of available phosphate.

Available nitrogen is generally low, and responses to nitrogen fertilizers have been obtained widely on cereal crops and cotton. In many soils peak availability occurs shortly after the start of the rains and thereafter declines, perhaps rapidly. Crop roots must compete against immobilization, which is particularly vigorous when C:N ratios are high after recent grass fallows, and against leaching. Rates of leaching vary widely, being very rapid in sandy profiles but much less so in Ferruginous Soils with a heavy textural B-horizon. Where leaching is rapid the timing of nitrogen fertilizer applications is critical.

The sulphur supply is delicately balanced. The only natural external source is a small contribution from the atmosphere which is usually less than the crop requirement, while losses from the soil-plant system through burning and, in some soils, by leaching may be considerable. However, sulphate is conserved by adsorption on the clay of textural B-horizons where these exist. Deficiency occurs in groundnuts and cotton and would be common but for the fact that these crops, being cash crops, are often fertilized with phosphate fertilizers containing sulphur.

The soil content of total potassium varies according to parent material, in general being low over sedimentary materials and moderately high over crystalline rocks; but because of the small exchange capacities levels of exchangeable potassium are usually low. In Q/I relationship terms, potassium intensity is high in most Ferruginous and Ferrallitic Soils, but quantities and buffer capacities are low or very low. On Vertisols, Eutrophic Brown and some Hydromorphic Soils the pattern is different, with low intensities but high values for the quantity parameter and buffer capacity. Crop deficiencies have been reported, but usually only after several years of intensive production, particularly on soils overlying certain sandstones low in potassium. However with potassium, more than with any other nutrient, the long-term balance depends very much on whether crop residues are returned to the soil.

Deficiencies of calcium and magnesium are rare but may be brought about by prolonged cropping of sandy soils. Lack of calcium is sometimes responsible for a low shelling percentage in groundnuts. Of the micronutrients, boron deficiency has been found in cotton and, less often, molybdenum in groundnuts.

(d) *Effects of traditional practices*

The majority of farmers today still rely very largely on traditional techniques. Where possible, land is allowed to revert to bush fallow after a few years of cropping. It is cleared by cutting and burning, and except in some drier areas crops are grown on mounds or ridges. Interplanting (intercropping or mixed cropping) predominates over sole cropping, and precedence is everywhere given to the production of food crops.

There has been little study of the effects of these practices on the soil, but some deductions can be made from other work. The main effects of fallowing are to accumulate nutrients, to increase soil organic matter and to suppress the weeds, pests and diseases of cultivation. Soil structure may be improved by grass fallows, but the effect is rather short-lived. The ratio of fallow years to cropping years thought necessary to maintain fertility varies widely according to the nature of the soil, the nature of the fallow vegetation and the level of fertility regarded as satisfactory. For the sake of the soil the longer the fallow the better, but even 3—4 year fallows may give a substantial benefit to crops in the following one or two seasons. Shorter fallows in combination with mineral fertilizers may also be effective. No planted fallow species has yet been found that substantially increases fallow efficiency.

Nutrient accumulation by the fallow is increased by the presence of deeper-rooted woody vegetation. Frequent bush fires, which tend to reduce tree cover, therefore cause not only immediate losses of nitrogen, sulphur and possibly other nutrients from the burned vegetation and litter but also a decline in fallow effectiveness. However, burning prior to cultivation, which is the only practicable way of clearing bush, is useful in releasing available phosphate to the following crops.

Organic manuring is widely practised but, because of the scarcity of manure, only at a very low level of application. Under these conditions the benefit it confers is restricted almost entirely to its nutrient contribution, although at high rates (rarely possible outside the experimental farm) it may increase soil organic matter and ameliorate physical and physico-chemical properties.

Intercropping has certain economic advantages: higher returns per man-hour, more efficient use of land (and hence tilling and weeding effort), more efficient use of the short growing season, and reductions in losses from diseases, insects and weeds; but effects on the soil have not been studied. They may include (i) more complete crop cover and therefore reduced erosion, (ii) better exploitation of the nutrient and moisture supply, both in depth and time, and (iii) a greater return of residues per unit area.

Under true subsistence farming there is considerable recycling of nutrients. Losses are restricted to volatilization of nitrogen and sulphur, largely through burning, and to a limited amount of leaching of nitrate and cations. Gains include small atmospheric contributions of nitrogen and sulphur, microbiological fixation of nitrogen, and the "pumping" of nutrients, particularly phosphate, from depth by the fallow vegetation. However, once cash crops are introduced there is likely to be an appreciable net loss of nutrients, especially phosphate, from the farmed area, unless suitable fertilizers are introduced at the same time.

(e) *Maintenance of fertility*

Using traditional techniques, prolonged cropping, without fallowing or fertilization, usually causes a rapid degradation in both physical and chemical properties of the soil and a severe decrease in productivity. To maintain fertility over a long period the soil resources must be carefully managed.

We conclude this summary by surveying research findings in the three main aspects of management: (i) control of water, (ii) maintenance of nutrient status, and (iii) maintenance of a good rooting medium.

(i) At many sites the balance between the provision of adequate moisture for crop growth, the prevention of erosion and the avoidance of surface waterlogging and excessive leaching is a delicate one. In general, the maintenance of optimal soil moisture has priority, but once this requirement has been met any excess water should be disposed of in the least erosive way.

Erosion control has been studied on runoff plots at a number of stations. Prevention of runoff, by absorption terraces, tied contour ridges or tied mounds, can increase yields in dry areas but may be harmful because of anaerobic conditions and even dangerous, due to the breaking of ridges and banks during heavy storms. Hence, in all except the driest areas, systems designed to reduce the rate of flow of runoff and to guide it safely into watercourses are usually more satisfactory. They include graded ridges and graded terraces. However, even in humid areas water is often a limiting factor at the beginning and end of the season, and management systems which permit an easy alternation between water conservation and water disposal according to time of season (e.g. by tying and untying ridges) have an advantage. Flat planting is safe only where the ground has been well levelled within a system of graded contour terraces.

Other measures include: (i) strip cropping along the contour, although for success this needs at least one "non-erosive crop" able to exert a braking effect on flowing water; (ii) narrow buffer strips of thick grass on the contour between wider strips of crops; and (iii) graded terraces, draining into grassed watercourses and separated by strong banks, which are complementary to graded ridges or flat planting.

Little information is available from southern savanna areas having a two-peak rainfall distribution. At some sites as little as 1000—1200 mm rain may be divided between two distinct seasons, each receiving perhaps a rather poorly distributed 400—600 mm, and the risk of yield reductions from water stress can be high (Bouchy, 1973). Moisture conservation measures may be as essential here as in semi-arid areas.

(ii) Most long-term fertility experiments have tested fertilizers in combination with different rotation, fallowing or manuring treatments, and under these conditions the maintenance of moderately high yields appears to be possible for ten years and probably longer. In francophone countries fertilization is often considered in terms of the whole rotation rather than the individual crop year (at least for phosphate applications), and a distinction is drawn between basal and maintenance fertilization. Basal fertilization aims to increase the fertility of the soil with one large dressing, its size depending on the degree of the existing deficiency, while maintenance dressings are applied to balance nutrient losses during cropping in order to maintain the fertility level established by the basal fertilizer.

So far, basal fertilization has been limited to correcting phosphate deficiency with rock phosphate, where this is available. The optimum size of application depends on soil type, ranging from 20—30 kg P/ha on many upland soils to as much as 400 kg P/ha on very strongly adsorbing Hydromorphic Soils. The principle of maintenance dressings is more difficult to apply, because there are few data on leaching losses and crop removal. In practice, most of the present fertilizer recommendations are based on the results of agronomic trials set up to determine the optimum economic response to fertilizer, but for the long-term improvement of the fertility of these well-weathered, poorly buffered soils the maintenance principle is an important one.

Choice between forms of fertilizer is usually very limited. Present trends are towards higher analysis materials, compounds, mixed fertilizers, and nitrogen sources less acidifying than ammonium sulphate. Any change from single to triple superphosphate must take into account the sulphur requirements of crops, particularly legumes. Also the increasing use of urea does not completely solve the problem of soil acidification, particularly as rates of application tend to increase. But the introduction of concentrated multinutrient fertilizers, by reducing transport costs and the labour of spreading, gives economic advantages.

(iii) Many savanna soils have high bulk density and low porosity. Significant correlations between crop yields, crop rooting depth and density and soil bulk density have been demonstrated in Senegal; and yield responses to ploughing (usually mouldboard) as compared with more superficial forms of cultivation have been obtained at many sites throughout West Africa. In general, yields increase with cultivation depth: tractor ploughing $>$ ox ploughing $>$ hand hoeing. On present evidence it seems that, although the savanna soil is often a poor rooting medium, it can be improved by ploughing.

Ploughing aids rooting by increasing the porosity and reducing mechanical impedance. The soil moisture regime is also involved. First, through improved infiltration and permeability, the soil may contain more available water at critical periods of growth; and, secondly, the greater root growth brought about by the loosened topsoil often leads to a deeper and more thorough exploitation of the whole moisture profile. Crops on ploughed soils have been reported to show greater drought resistance.

The durability of improved porosity and lowered bulk density following ploughing is very variable. In some cases the improvement lasts only a few weeks, although this may be sufficient to give the crop a significantly better start. Among the factors affecting the preservation of improved structure into the second season after ploughing are: (a) soil texture, (b) intensity of rainfall in the first season, (c) condition of initial cultivation, e.g. cloddy versus fine tilth, (d) cropping practice and harvesting operations (the harvesting of groundnuts is a destructive process), and (e) nature of crop, its rooting habit and the degree of soil cover it gives.

In addition, there is evidence that structural improvements from ploughing are more long-lasting if organic materials, e.g. crop residues, herbaceous fallow or green manure crops, are incorporated in the soil by the ploughing operation. This necessitates ploughing at the end of the season, because it is important that the organic material be partially decomposed before the next crop is planted. Incorporation of crop residues is therefore possible only after a short-season crop. Further, where the season is particularly short, as in northern and central Senegal, ploughing at the beginning of the season delays planting to an unacceptable degree. Hence, ploughing is possible only if one season per rotation is set aside for it, and a grass fallow or green manure crop is grown, which can be ploughed in at the right time, just before the end of the rains. In these circumstances there appears to be some justification for growing a green manure crop, although it may be difficult to persuade farmers of the benefits. Better still would be an economically productive "fallow" that left enough residues for incorporation before the end of the season.

2 THE MAINTENANCE OF FERTILITY UNDER LOW INTENSITY FARMING

(a) *Control of water*

There is little evidence that traditional agriculture embraced any techniques to ensure safe, non-erosive runoff. Terracing was used in some hill areas, but almost certainly this was a measure intended primarily to give an adequate depth of soil for crop rooting. Similar remarks apply to mounds and ridges. The latter may still sometimes be seen aligned down steep slopes, where they manifestly encourage rapid gullying. The abundance of unoccupied land in the past clearly gave little incentive for the adoption of conservation practices, but with increasing populations and the introduction of costly technical innovations an education programme in soil conservation is needed. Except where there is an overriding need for water conservation, open ridges slightly off the contour are likely to give greatest protection, with grass buffer strips between fields to prevent large-scale movements of soil.

In contrast, the control of soil moisture through small-scale irrigation systems has long been practised in the savanna. Monod and Toupet (1961), tracing the history of land use in the Sahel zone, distinguished two systems of cultivation, that utilizing residual soil moisture following the impounding of floodwater by dams thrown across wadis, and that using irrigation, usually by shaduf. While the former is confined to the valleys of semi-desert areas and has very limited development potential, shaduf irrigation of dry season vegetable gardens is already common in many parts of the region south of the Sahel zone. Replacement of the laborious shaduf with a diesel pump increases the area that can be watered, and the growth of urban markets and the establishment of canning factories make such an enterprise highly profitable.

Unfortunately there has been little study of the fertility of these irrigated garden soils. Fertility has usually been maintained by fallowing and manuring with domestic refuse, but with the sale of goods for cash will come the use of mineral fertilizers. It seems certain from work on vegetables elsewhere that optimum fertilizer applications will be high.

(b) *Maintenance of nutrient status*

The most important factor in the maintenance of soil nutritional levels in the past was the bush fallow, but it is on the fallow, and particularly on its duration, that the pressure of increasing population and the spread of cash crops is now felt. If land is short, the area under fallow declines, and fertilizers are used instead.

However, there are dangers associated with the use of mineral fertilizers on these soils, which need some emphasis. First, the fertilizers commonly used supply only two or three nutrients, and since release from mineral reserves is very slow, the available levels of other nutrients decrease as cultivation continues. Secondly, the fertilizer dressings recommended to farmers are usually those that have been established as the most economic in annual experiments, and their repeated application under more continuous cropping conditions can lead to a soil impoverishment even of those nutrients that are applied. Many farmers now need recommendations based on longer-term experiments if they are to maintain the fertility and productivity of their land.

Similar points have been made recently by three authors working independently in different parts of the savanna. Richard (1970) argued that the fertilizer formulae at present supplied to farmers should be regarded only as a temporary solution to the problem of increasing yields, because they did not replace all the exported nutrients; Pieri (1973) commented that this type of fertilization concealed behind an apparent economic profitability the fact that the soils were being impoverished; and Bouchy (1973) wrote that not only is fertilizer application limited to profitable crops, but even these are supplied with only the immediately profit-bringing fraction of the necessary nutrients, which results in an undesirable exploitation of the natural reserves of the soil and a reduction in the agronomic value of the land.

The nature of the recommendations to be made to farmers has been considered in detail by agronomists at Bambey. From collected data on yield levels, nutrient export in the economic yield, prices of produce and costs of fertilizers and management operations, Tourte (1971) and his colleagues produced balance sheets both of soil nutrients and farm economics under different mineral fertilizer regimes for three groundnut-growing zones of Senegal. The nutrient balance sheet and other relevant data are summarized in Table 74.

From this work it was concluded that cropping without fertilizers (F_0) brought a catastrophic degradation of soil resources and a grave danger of acidification in all three zones; but also that the current fertilizer recommendations (F_1) made to farmers not possessing ox or tractor-drawn equipment brought a decline in soil reserves even of some of those nutrients included in the fertilizer. Potassium was undersupplied, and the calcium requirement barely met. Since these calculations took no account of other mechanisms of loss from the soil, e.g. erosion and leaching, it seems likely that available levels of both cations would steadily decline.

Rotation fertilization (F_2) supplied sufficient potassium and, from the heavy dressings of rock phosphate in the fallow year, an excess of phosphate and calcium. However, because this latter system involves ploughing to incorporate the rock phosphate and the fallow vegetation, as yet it is beyond the reach of many farmers. Economically, it is justified over much of the Sine Saloum and Eastern zones of Senegal but, because of lower yield levels, only in the most favourable areas of the Northern zone. Here it is justified largely in terms of soil improvement.

This study was limited to only four nutrients: nitrogen (not included in Table 74), phosphorus, potassium and calcium; but others, magnesium, sulphur and some of the micronutrients, are also important. For maintenance fertilization to be put on a sound basis there is an urgent need for more detailed nutrient balance studies of all savanna farming systems.

(c) *Maintenance of a good rooting medium*

Many farmers construct mounds or ridges before planting and so create a deepened rooting zone of well-drained topsoil. Berger (1964), in Ivory Coast, observed that it was important to build the ridge or mound over loosened soil. Scraping up the soil on to a hard surface limited tap-root penetration of crops like cotton. In yam mounds the roots readily exploited the soil of the mound, but tuber elongation stopped when it reached the hard soil beneath the mound. It seems likely that this is the reason for the very large ridges (up to 70 cm high) constructed for yams in parts of northern Nigeria. However, the greater part of the farmed area, ridged or unridged, is cultivated to only a shallow depth. New ridges are often made by scraping soil over an undisturbed surface, and only about half the area is actually broken by the hoe. This allows rapid land preparation at a time of peak

TABLE 74

Nutrient balance sheet for three fertilizer regimes in three zones of Senegal (Tourte, 1971).

	Fertilizer level	Nutrient:	*Northern groundnut basin*			*Sine Saloum*			*Eastern Senegal*		
			P	*K*	*Ca*	*P*	*K*	*Ca*	*P*	*K*	*Ca*
Crop removal and fertilizer application of nutrients (kg/ha)	F_0	Removed	—10	—42	—23	—14	—63	—44	—15	—49	—26
	F_1	Removed	—12	—51	—34	—17	—77	—51	—21	—67	—36
		Applied	+21	+33	+32	+30	+33	+46	+26	+21	+38
		Difference	+ 9	—18	— 2	+13	—44	— 5	+ 5	—46	+ 2
	F_2	Removed	—13	—53	—34	—20	—91	—61	—25	—77	—39
		Applied	+83	+91	+175	+116	+91	+245	+119	+76	+249
		Difference	+70	+38	+141	+96	0	+184	+94	— 1	+210
Yields assumed in above balance sheet (kg/ha)		Crop*	G'nuts	Millet	G'nuts	G'nuts	Sorghum	G'nuts	Cotton	Sorghum	G'nuts
		F_0	950	430	960	1560	1140	1390	1370	1290	1550
		F_1	1090	700	1170	1790	1710	1710	1650	1950	2140
		F_2	1140	1020	2070	2080	2330	1980	2080	2870	2270

*Rotation in each case is of four years, one-year herbaceous fallow followed by 3 years cropping.
F_0 no fertilizer applied. F_1 current local fertilizer recommendations for hoe farming.
F_2 rotation fertilization (see Chapter 10): rock P on fallow, NK on groundnuts and cotton, N on cereals.
F_0 and F_1 applied to fields cultivated with horse-pulled hoe; F_2 with ox-ploughing once per rotation, incorporating fallow.

labour demand. Where the only available implement is the hoe, deeper cultivation of large field areas cannot be attempted.

Moreover, there are the constraints of the cropping cycle. For example, in parts of northern Nigeria cotton is often followed by a millet-sorghum intercrop. Since the cotton remains on the land until well into the dry season, soil preparation is not practicable until the rains restart; but the millet, a short-season variety, is highly valued as a hunger-breaker, so that a high premium is placed on early planting. To achieve this, soil preparation is omitted, and the seeds are pressed into slightly loosened soil immediately after the first heavy rain.

From the evidence in Chapter 11 it seems probable that high soil bulk density in the rooting zone is often a limitation on the productivity of farmers' fields, but as long as the hoe remains the only implement available there is little that can be done to tackle the problem. Tractor-hire units are now operating in a few areas, and this facility seems likely to expand. However, it is timely here to repeat the distinction drawn by Charreau and Nicou (1971c) between occasional deep, machine cultivations and frequent superficial soil working. While the former improves rooting and percolation, there is a high risk that the latter will cause a deterioration in soil structure and aggravate erosion. Thus unskilful mechanization may do more harm than good.

Work in Senegal and Upper Volta has shown that ox-ploughing, though less effective than tractor-ploughing, usually increases yields over those from hoe cultivations. In addition to giving a greater depth of loosened soil, it permits both the incorporation of organic residues and an earlier completion of land preparation and so an earlier sowing. Government incentive schemes to expand the ownership of oxen are therefore well justified, provided they are supported by effective training and advisory services.

(d) *Crop residues*

The way in which crop residues are disposed of may affect both the chemical and physical properties of the soil; but in the context of low intensity farming discussion is hampered by the lack of reliable information as to what farmers' current practices are. It seems to be generally true that the haulm (or hay) of the main legume crops, groundnuts and cowpeas, is almost completely used as livestock fodder, so that the question of burning or incorporation does not arise; but the residues of other crops are rarely gathered for this purpose, although they may be grazed *in situ*. Their importance in this respect almost certainly increases northwards with the decrease of good dry-season grazing.

Van Raay and De Leeuw (1970) examined the yield of crop residues and their fodder potential in Katsina Province, northern Nigeria. Millet residues were little used because, following the early harvest, they were usually badly decomposed before livestock could be allowed access to the fields (which carried mixed crops). Sorghum residues, however, were an important grazing resource, and between 24 and 69 per cent of samples presented to experienced herdsmen were judged grazeable. The proportion varied inversely with the quality of the crop, as the heavy, woody stems of high-yielding plants are not eaten. With cotton, grazing was limited to leaves, bolls and surface litter.

Stems of millet and sorghum are widely used for fencing. They may be lightly burned first, on the field, to harden them and remove the remaining leaves. But farmers carry off only the materials they need, and those that are not needed are either grazed, left to decompose, or, if they are in the way or liable to harbour pests, burned.

On this evidence, crop nutrients are moderately well conserved by existing practices. Even where residues are a vital source of animal feed utilization is incomplete, and in any case some return through the animal to the soil will occur. One unknown factor, however, is the extent to which decomposition of residues by termite activity involves the loss of nutrients out of the cropped area. No research appears to have been undertaken on this topic.

Apart from a certain mulching action, crop residues are likely to have little effect on soil physical properties where the hoe is the only implement available for cultivation. But, as has been described in Chapter 11, where power (ox or tractor) and time are available to incorporate the residues into the soil, the result is likely to be improved structure and hence better crop rooting for the next one, two and even sometimes three seasons. One may add in this connection that some termite species are regarded as beneficial, their activities producing a "breadcrumb" or pumicestone structure in the soil (Charreau and Nicou, 1971c).

However, even where tractive power is available, there are problems to be overcome before residue incorporation can be universally recommended. First, residues should be incorporated at the end of the rains, but this is often prevented by the length of the growth cycle. Postponement to the start of the next season may delay planting and impair the growth of the next crop, particularly if no extra nitrogenous fertilizer is applied. The severity of this problem depends on the length of the growing season and the sequence of crops in the rotation. Where ploughing with incorporation is accorded sufficient importance, rotations may be rearranged to fit in with them. This is a large part of the justification for the fallow (or green-manuring) year in the IRAT system in Senegal (Chapter 11).

A second problem concerns the nature of the residues. The thick stems of sorghum, maize and, to a lesser extent, millet need chopping before incorporation. This requires power. Alternatively, the stems can be laid in the furrows and buried by breaking the ridge over them, but this appears not to have been tested experimentally. A third possibility is to develop crop varieties with shorter, thinner stems, which can be incorporated without chopping. This is currently being investigated with millet at Bambey.

3 THE MAINTENANCE OF FERTILITY UNDER HIGH INTENSITY FARMING

We have said that the second aim of research should be to devise high-productivity systems of agriculture which utilize the land on a stable, permanent basis. The purpose is twofold:—

(i) To demonstrate and explore the potential of the agricultural environment and provide a target for the improvement of low intensity farming.

(ii) To provide advice to governments and private individuals wishing to make a capital-intensive investment in agriculture. At present there is little demand for such advice, but prices of agricultural commodities, particularly foodstuffs, are rising fast, and high-productivity agriculture could soon become highly profitable. Already, some governments are spending large sums of money on new irrigation schemes, even though, unfortunately, there are few research results to guide them.

(a) *Control of water*

Much of the work reviewed in Chapter 11 on runoff and erosion control concerned land within potentially high productivity systems, and there is no need for further discussion. In any case, land conservation systems have been described by Hudson (1971). It is sufficient to say that any large-scale agricultural development in the savanna in which conservation measures are neglected will almost certainly end in disaster.

The limitation imposed on agriculture by the long dry season and the recent serious drought in the Sahel zone make large-scale irrigation appear a very attractive proposition, and many such projects have been proposed in the last few years. A few large-scale irrigation schemes have already existed for several decades, particularly for the production of rice, cotton and sugar-cane. The biggest area is in south-western Mali, where by 1962 about 50,000 ha of the inland Niger delta were irrigated for cotton and rice production under the Office du Niger and a further 150,000 ha in the same area were included in smaller schemes to grow food crops for local needs and export (Morgan and Pugh, 1969, pp 645—652). Other projects have included irrigated rice at Richard-Toll in the Senegal River valley and the creation of polders to grow wheat and maize at Bol (Chad Republic) on the edge of Lake Chad (Guichard *et al.*, 1959).

For the future, there is good quality water available from the perennial rivers, the Niger, Senegal and Volta, and their tributaries, and from a few smaller rivers. It may also be possible to make more use of groundwater supplies, particularly those of the sedimentary basins underlying much of the Sahel zone, but data on recharge rates will be needed, and not all groundwater is of sufficiently high quality.

However, it is arguable that where money is short it may often be better spent in subsidizing inputs of fertilizer, improved seed, pest-control equipment and trained advisory manpower for dryland farming, for which there are the results of more than fifty years research. Insufficient knowledge of the agronomic and pedological conditions and social problems were among the initial difficulties encountered by the inland Niger delta scheme, and outside the few well-established sites very little research effort has been made towards avoiding these difficulties in future developments.

For large-scale irrigation to be profitable, farming standards must be high, efficient use must be made of the water, and regular checks on groundwater levels and salinity are essential. As with all other innovations, irrigation should be developed only by taking into account all aspects of crop production, including the selection of suitable sites, the control of weeds and disease, and, not least, the long-term control of fertility.

Undoubtedly there is a high potential for irrigation in the savanna, but it is advisable only after pilot projects have tackled the pedological, agronomic and social problems that arise from it.

(b) *Maintenance of nutrient status*

As the nutrient status of most savanna soils is naturally low, the establishment of a high-productivity system of agriculture requires an initial build-up of fertility. The discussion of basal fertilization in Chapter 10 is relevant here. However, in practice, basal fertilization is limited almost entirely to phosphate applications, usually as rock phosphate.

Within limits determined by the soil type, adsorption reactions hold applied phosphate in the surface soil. The basal application is thought of as blocking the adsorption sites—or as satisfying the fixation capacity of the soil—in order that subsequent maintenance dressings may be fully effective; but where the adsorbed forms remain available, the basal application is also useful in increasing the size of the labile pool. Except where the applications are very heavy or the soils very sandy, leaching losses are almost negligible.

One may perhaps see sulphur in similar terms (Chapter 8). Where there is a textural B-horizon to adsorb sulphate, high initial applications of sulphur are largely conserved in available form for subsequent crops. One drawback is that the adsorbing horizons may be too deep for optimum recovery by shallow-rooted crops.

The concept of basal fertilization is inappropriate to nitrogen. Its annual mineralization depends largely on the organic matter content of the soil, and, while there is some scope for build-up through careful management, it is very doubtful whether under conditions of continuous arable farming, native soil mineral nitrogen will ever be sufficient to support a heavy cereal crop. The inclusion of an effective grain legume in the rotation may help, but highest yields will need supplementary dressings of nitrogen fertilizers.

For the other nutrients, calcium, magnesium and potassium, the application of the concept of basal fertilization is severely restricted by the low exchange capacity of the soil. All that can be done is to maintain the organic matter at as high level as possible and the soil pH in the range 6·0—7·0.

As cropping continues, regular maintenance dressings of all nutrients will be needed; but the amounts to be applied need careful calculation. Large applications of phosphate may have a useful residual effect, but excess nitrogen, apart from being uneconomic and possibly depressing yields, will promote cation leaching and acidification; and an excess of one cation may prejudice uptake of another. One factor of importance is the effect of the return of crop residues, burnt or unburnt, on the balance of cations in the topsoil. Most residues are high in potassium, and the annual return to the surface soil of perhaps 100—150 kg K/ha taken up from the whole profile can lead to this cation occupying an unfavourably high proportion of the exchange sites. Except in soils very poorly supplied in potassium, optimum maintenance dressings may need to contain relatively more calcium and magnesium than potassium.

A further point is that when a fertilizer like potassium chloride is applied, the subsequent downward leaching of the chloride anions necessarily involves the similar movement of an equivalent number of cations. If the potassium is retained, then other cations are displaced and lost. On the other hand, if potassium is applied as a phosphate, adsorption will hold the phosphate anions in the topsoil, and there will be no loss of cations. The same may be true when, applying potassium as potassium nitrate, the nitrate ion is efficiently intercepted by plant roots. Such effects, of little importance in well-buffered soils, may need to be considered when calculating nutrient balance sheets and estimating fertilizer dressings for continuous, intensive cropping of savanna soils. In this context, it is highly desirable to know their chemical composition before multinutrient compound fertilizers are introduced.

Finally, the question whether continuous cropping will be possible under all conditions without the use of "resting" crops or bush fallows cannot yet be answered. The evidence from the experimental station is that with the judicious use of fertilizers (and other inputs) high yields can be produced for many years. But to the farmer, costs of inputs and the need for a secure food supply are of overriding importance; and there are wide areas where there has so far been little experimental work. Studies of long-term trends of fertility under different farming systems are therefore essential.

(c) *Maintenance of a good rooting medium*

The work reviewed in Chapter 11 showed that the improved soil structure resulting from cultivation of the soil to a depth greater than that usually achieved by hand tillage usually increased crop growth, and that this improvement could be given greater permanence by incorporating organic materials into the soil during ploughing. Under high intensity farming the necessary traction for these operations may be assumed to be available. The main difficulty would be timing.

In Senegal it is recommended that this problem be overcome by including in the rotation at least one year's herbaceous fallow or a green-manure crop, as discussed above; but in an intensive system one would not wish to have unproductive years in the rotation. Alternatively, it has been suggested that the fallow year be used to grow a fodder crop, with the regrowth ploughed in as green manure.

This latter suggestion is a radical one for savanna agriculture, for all other innovations and proposals appear to accept the traditional separation of crop-raising and stock-raising activities. There is no clear practical reason why this should be so. Reference is often made to the great benefits for the soil to be obtained from the use of farmyard manure; animal traction is badly needed; we know that animals eat some crop residues; and there are examples, e.g. at Shika, Nigeria, of cattle and crops being integrated in a successful farming enterprise. But there appears to have been no comprehensive study of the maintenance of soil fertility in a mixed farming system.

We therefore refer here to the maintenance of fertility in Australian wheat soils (Greenland, 1971), which we believe has considerable relevance to the savanna situation. Although the climate is less harsh, the surface soils have high silt or fine-sand contents and under cultivation tend to develop a structure unsuited to plant growth; and there is strong evidence of a link between structural conditions and wheat yields. Current practice is to alternate several years of cropping to wheat with several years under legume or grass/legume pastures. The effect of the pasture phase is twofold; first, it accumulates readily mineralizable organic nitrogen, and amounts released during the arable phase are sufficient to produce at least moderate wheat yields. Secondly, important physical changes occur, for example, increases in porosity, aggregate and microaggregate stability and water retention; and Greenland concluded that pastures are essential for the maintenance of the physical condition of most Australian wheat soils with high contents of silt and fine sand.

In the savanna, stock-raising is usually nomadic, and under ideal conditions this provides some grazing all the year round. Nevertheless, it is a highly inefficient activity in terms of land use, with poor quality animals dependent on large areas of poor quality bush, and already the pastoralist is competing with the cultivator for land. The soils, many with high fine-sand contents, are low in available nitrogen and have a poor structure that deteriorates on cropping. The old remedy, the bush fallow, has a limited future in the face of rapidly increasing populations. The Australian example points to a very promising field of co-operative research between agronomists, soil scientists and animal husbandry specialists, with a highly productive mixed farming system suitable for savanna conditions as the ultimate goal.

4 CONCLUSIONS

The three main elements essential to the maintenance and improvement of fertility in savanna soils may be summarized as:

1. Maintenance of the existing stock of topsoil, through carefully planned erosion control.

2. Maintenance of the existing stock of crop nutrients, by conserving those already present (returning residues, ash, manure, etc) and by replacing through fertilizer use those known to have been lost by leaching and crop export. (Before this maintenance phase is reached an initial build-up, especially of phosphate, will be necessary in nearly all soils.)

3. Improvement of the soil as a physical medium for crop rooting, by appropriate tillage operations and moisture control.

We place firm emphasis on maintenance and conservation at a time when social and economic pressures encourage exploitation, for we believe it is the responsibility of the scientist and extension worker to ensure that the innovations they introduce to improve the present lot of the farmer and the community do not harm the soil nor prejudice its more efficient utilization in the future. Research has already indicated methods of soil management which will increase yields and yet maintain or improve fertility. Many problems remain, but perhaps the greatest is the development of research findings into practicable recommendations that the farmer is able and willing to apply. Due weight must be given to this vital area of study, by soil scientists as well as by economists and sociologists, if the high potential productivity of the savanna is to be achieved.

LOCATION OF PLACES REFERRED TO IN TEXT AND TABLES.

Place	*Country*	**Location*	
Adiopodoume	Ivory Coast	5.19 N	4.01 W
Allokoto	Niger	14.30 N	5.30 E
Bambari	Central African Republic	5.40 N	20.37 E
Bawku	Ghana	11.05 N	0.11 W
Bambey	Senegal	14.40 N	16.28 W
Birao	Central African Republic	9.30 N	21.30 E
Bol	Chad	13.27 N	14.40 E
Bobo Dioulasso	Upper Volta	11.11 N	4.18 W
Bouake	Ivory Coast	7.42 N	5.00 W
Boukombe	Dahomey	10.13 N	1.09 E
Boulel	Senegal	14.20 N	15.34 W
Darou	Senegal	14.09 N	16.08 W
Diankancounda Ogueil	Senegal	13.00 N	14.30 W
Diebougou	Upper Volta	11.00 N	3.12 W
Dieri	Mauritania	15.05 N	12.34 W
Dilbini	Chad	12.05 N	17.04 E
Djibelor	Senegal	12.35 N	16.20 W
Ejura	Ghana	7.23 N	1.15 W
Farako Ba	Upper Volta	10.45 N	6.50 W
Ferkessedougou	Ivory Coast	9.30 N	5.10 W
Fort Lamy (Ndjamena)	Chad	12.10 N	14.59 E
Grimari	Central African Republic	5.43 N	20.06 E
Guetele	Cameroun	10.53 N	13.54 E
Ibadan	Nigeria	7.23 N	3.56 E
Ilorin	Nigeria	8.32 N	4.34 E
Jebba	Nigeria	9.11 N	4.49 E
Jos	Nigeria	9.54 N	8.53 E
Kaduna	Nigeria	10.28 N	7.25 E
Kaffrine	Senegal	14.08 N	15.34 W
Kano	Nigeria	12.00 N	8.31 E
Katsina	Nigeria	13.00 N	7.32 E
Kontagora	Nigeria	10.24 N	5.22 E
Koupela	Upper Volta	12.07 N	0.21 W
Kwadaso	Ghana	6.45 N	1.35 W
Louga	Senegal	15.37 N	16.13 W
Man	Ivory Coast	7.31 N	7.37 W
Matam	Senegal	15.40 N	13.18 W
Mokwa	Nigeria	9.19 N	5.00 E
Mopti	Mali	14.29 N	4.10 W
Mouka	Central African Republic	7.20 N	22.00 E
Navrongo	Ghana	10.51 N	1.03 W
Niamey	Niger	13.32 N	2.05 E
Niangoloko	Upper Volta	10.20 N	4.50 W
Niari Valley	Congo Republic	4.15 S	13.30 E
Nioro-du-Rip	Senegal	13.45 N	15.48 W
Nyankpala	Ghana	9.24 N	1.00 W
Oursi	Upper Volta	14.42 N	0.29 W
Patar	Senegal	14.33 N	16.24 W
Richard-Toll	Senegal	16.25 N	15.42 W
St. Louis	Senegal	16.01 N	16.30 W
Samaru	Nigeria	11.11 N	7.38 E
Sare Bidji	Senegal	13.00 N	14.30 W
Saria	Upper Volta	12.15 N	2.23 W
Sefa	Senegal	13.10 N	15.30 W

Place	*Country*	**Location*	
Shika	Nigeria	11.14 N	7.35 E
Sinthiou-Maleme	Senegal	14.03 N	15.18 W
Sokoto	Nigeria	13.02 N	5.15 E
Tambacounda	Senegal	13.45 N	13.40 W
Tarna	Niger	13.29 N	7.10 E
Thies	Senegal	14.49 N	16.52 W
Tivaouane	Senegal	14.57 N	16.45 W
Tougou	Upper Volta	13.30 N	2.20 W
Yandev	Nigeria	7.22 N	9.01 E
Zaria	Nigeria	11.01 N	7.44 E
Zinder	Niger	13.46 N	8.58 E

*Note: where co-ordinates were not given in original reference, they have been estimated from descriptive material and maps and are therefore only approximate.

REFERENCES

ABDULLAHI, A. 1971. Monocropping/rotation trial—Samaru. Unpublished internal report, Inst. Agric. Res., Samaru, Nigeria.

ACQUAYE, D.K. 1963. Some significance of soil organic phosphorus mineralization in the phosphorus nutrition of cocoa in Ghana. *Pl. Soil*, **19**, 65-80.

ACQUAYE, D.K., MACLEAN, A.J. AND RICE, H.M. 1967. Potential and capacity of potassium in some representative soils of Ghana. *Soil Sci.*, **103**, 79-89.

ACQUAYE, D.K. AND OTENG, J.W. 1972. Factors influencing the status of phosphorus in surface soils of Ghana. *Ghana J. agric. Sci.*, **5**, 221-228.

ADEJUWON, J.O. 1969. Vegetation zonation and the distribution ranges of savanna trees in Nigeria. *Niger. geog. J.*, **12**, 125-133.

ADU, S.V. 1963. The soils of Tono state farm. Tech. Rep. No. 58, Soil and Land Use Survey, Agric. Res. Inst., Kumasi, Ghana.

ADU, S.V. 1972. Eroded savanna soils of the Navrongo-Bawku area, northern Ghana. *Ghana J. agric. Sci.*, **5**, 3-12.

AGRICULTURAL DEPT., NORTHERN NIGERIA. 1953. Internal Memo.

AHN, P. M. 1968. The effects of large-scale mechanized agriculture on the physical properties of West African soils. *Ghana J. agric. Sci.*, **1**, 35-40.

AHN, P. M. 1970. *West African Soils*. Oxford Univ. Press.

AKEHURST, B. C. AND SREEDHARAN, A. 1965. Time of planting—a brief review of experimental work in Tanganyika 1956-62. *E. Afr. agric. For. J.*, **30**, 189-201.

ALEXANDER, M. 1965. Nitrification, pp. 307-343 in *Soil Nitrogen*, ed. Bartholomew, W. V. and Clark, F. E. Am. Soc. Agron., Madison, Wisconsin.

ALLAN, A. Y. 1968. Maize diamonds: some valuable results from the district husbandry trials in 1966. *The Kenya Farmer*, Jan. 1968.

ALLAN, A. Y. 1971. *The influence of agronomic factors on maize yields in Western Kenya with special reference to time of planting*. Ph. D. Thesis, Univ. East Africa, Nairobi.

ALLAN, W. 1965. *The African Husbandman*. Oliver and Boyd, Edinburgh and London.

ALLISON, F. E. AND STERLING, L. D. 1949. Nitrogen formation from soil organic matter in relation to total nitrogen and cropping practices. *Soil Sci.*, **67**, 239-252.

AMAGOU, V. 1965. Considerations sur la fertilite des sols et la fertilisation des cultures en Cote d'Ivoire. *Proc. 2nd meeting on soil fertility and fertilizer use in W. Africa*, **2**, 321-35.

ANDREWS, D. J. 1968. Wheat cultivation and research in Nigeria. *Niger. agric. J.*, **5**, 67-72.

ANDREWS, D. J. 1970. Relay and intercropping with sorghum at Samaru. *Proc. Seminar on Intercropping, Ford Foundation/IITA/IRAT, Ibadan*, 1970.

ANDREWS, D. J. 1974. Responses of sorghum varieties to intercropping. *Expl. Agric.*, **10**, 57-63.

ANON. 1959. Ann. Rpt. Dept. Agric. Northern Region of Nigeria, 1958-9.

ANON. 1967. Colloque sur la fertilite des sols tropicaux, Tananarive.

ARNON, I. 1972. *Crop Production in Dry Regions* (2 vols). Leonard Hill, London.

ASHRIF, M. I. AND THORNTON, I. 1965. Effects of grass mulch on groundnuts in the Gambia. *Expl. Agric.*, **1**, 145-152.

AUBERT, G. 1965. Classification des sols utilisee par la Section de Pedologie de l'Orstom. *Cah. ORSTOM, Ser. pedol.*, **3**, 269-288.

AUBERT, G. AND TAVERNIER, R. 1972. Soil Survey, pp. 17-44 in *Soils of the Humid Tropics*. Natl. Acad. Sci., Washington.

AUBREVILLE, A. 1949. *Climats, forets et desertifications de l'Afrique tropicale*. Societe d'Editions geographiques, maritimes et coloniales, Paris.

AUBREVILLE, A. ET AL. 1959. *Vegetation Map of Africa South of the Tropic of Cancer*. Explanatory notes by Keay, R. W. J. Oxford Univ. Press.

AUDRY, P. 1961. *Etude pedologique du cercle de Guidimaka (Mauritanie)*, Vol. I. ORSTOM, Hann, Dakar.

AUDRY, P. 1967. Observations sur le regime hydrique compare d'un sol ferrugineux tropical faiblement lessive sous savane et sous culture. *Colloque sur la fertilite des sols tropicaux, Tananarive*, **2**, 1591-1614.

AYLEN, D. 1939. *Soil and Water Conservation,* Pt. II. Rhodesian Govt. Printer, Salisbury, Rhodesia.

BACHE, B. W. 1965. The harmful effects of ammonium sulphate on fine sandy soils at Samaru. *Proc. 2nd meeting on soil fertility and fertilizer use in W. Africa,* **1,** 15-20.

BACHE, B. W. AND HEATHCOTE, R. G. 1969. Long-term effects of fertilizers and manure on soil and leaves of cotton in Nigeria. *Expl. Agric.,* **5,** 241-247.

BACHE, B. W. AND ROGERS, N. E. 1970. Soil phosphate values in relation to phosphate supply to plants from some Nigerian soils. *J. agric. Sci., Camb.,* **74,** 383-390.

BACHELIER, G., CURIS, M. AND MARTIN, D. 1956. The savanna soils of the South Camerouns. *Proc. CCTA/CSA Conf. Inter. Afric. ocident, Sao Tome, VI Session,* **2,** 19-36.

BAKER, R. M., LEEFERS, C. L., DE ROSAYRO, R. A., VAN DER SLUYS, D. H. AND TAYLOR, B. W. 1965. *Land use survey of the Western Region of Nigeria,* Vol. II. FAO Report.

BALANDREAU, J. AND VILLEMIN, G. 1973. Fixation biologique de l'azote moleculaire en Savane de Lamto (basse Cote d'Ivoire). Resultats preliminaires. *Rev. Ecol. Biol. Sol,* **10** ,25-33.

BANG, N. C. AND OLIVER, R. 1971. Contribution a l'etude du statut phosphorique des sol Madagascar. *Terr. Malgache,* **12,** 179-193 (quoted by Pichot and Roche, 1972).

BARBER, R. G. AND ROWELL, D. L. 1972. Charge distribution and the cation exchange capacity of an iron-rich kaolinitic soil. *J. Soil Sci.,* **23,** 135-146.

BARLEY, K. P. AND GRAECEN, E. L. 1967. Mechanical resistance as a soil factor influencing the growth of roots and underground shoots. *Adv. Agron.,* **19,** 1-43.

BARROW, N. J. 1961. Phosphorus in soil organic matter. *Soils Fertil.,* **24,** 169-173.

BARROW, N. J. 1967. Studies on extraction and on availability to plants of adsorbed plus soluble sulphate. *Soil Sci.,* **104,** 242-249.

BARROW, N. J. 1973. Relationship between a soil's ability to adsorb phosphate and the residual effectiveness of superphosphate. *Aust. J. Soil Res.,* **11,** 57-63.

BARTHOLOMEW, W. V. 1965. Mineralization and immobilization of nitrogen in the decomposition of plant and animal residues, pp. 285-306 in *Soil Nitrogen,* ed. Bartholomew, W. V. and Clark, F. E. Am. Soc. Agron., Madison, Wisconsin.

BARTLETT, H. H. 1957. *Fire in relation to primitive agriculture and grazing in the tropics,* Vol. II. Ann Arbor, Michigan.

BERGER, J. M. 1964. Profils culturaux dans le centre de la Cote d'Ivoire. *Cah. ORSTOM, Ser. pedol.,* **2,** 1, 41-69.

BERTRAND, R. 1967a. Etude de l'erosion hydrique et de la conservation des eaux et du sol en pays Baoule (Cote d'Ivoire). *Colloque sur la fertilite des sols tropicaux, Tananarive,* 2, 1281-1295.

BERTRAND, R. 1967b. L'erosion hydrique. Nature et evolution des materiaux enleves. Relations et consequences sur le sol erode. *Colloque sur la fertilite des sols tropicaux, Tananarive,* **2,** 1296-1301.

BERTRAND, R. 1971. Reponse de l'enracinement du riz de plateau aux caracteres physiques et chimiques du sol. *Agron. trop., Paris,* **26,** 376-386.

BERTRAND, R., NABOS, J. AND VICAIRE, R. 1972. Exportations minerales par le mil et l'arachide consequences sur la definition d'une fumure d'entretien d'un sol Ferrugineux Tropical developpe sur materiau eolian a Tarna (Niger). *Agron. trop., Paris,* **27,** 1287-1303.

BEYE, G. 1972. Etude de sols tropicaux: acidification, toxicite et amendment. *Paper read at Ford Foundation/IRAT/IITA symposium on Tropical Soil Research, Ibadan,* May 1972.

BHAT, K. K. S. 1970. *Contribution a l'Etude de la Dynamique du Phosphore dans les Sols Tropicaux et de ses Consequences sur l'Alimentation Phosphorique des Vegetaux.* These Doct., Fac. Sci., Univ. Paris.

BHAT, K. K. S. AND BOUYER, S. 1968. Influence de la matiere organique sur le phosphore isotopiquement diluable dans quelques types de sols tropicaux. *Proc. symp. isotopes and radiation in soil organic matter studies,* 299-313. Int. Atomic Energy Agency, Vienna.

BIRCH, H. F. 1958. The effect of soil drying on humus decomposition and nitrogen availability. *Pl. Soil,* **10,** 9-31.

BIRCH, H. F. 1959. Further observations on humus decomposition and nitrification. *Pl. Soil,* **11,** 262-292.

BIRCH, H. F. 1960. Nitrification in soils after different periods of dryness. *Pl. Soil,* **12,** 81-96.
BIRCH, H. F. 1961. Phosphorus transformations during plant decomposition. *Pl.* Soil, **15,** 347-366.
BIRCH, H. F. AND EMECHEBE, A. M. 1966. The effect of soil drying on millet. *Pl. Soil,* **24,** 333-335.
BIRCH, H. F. AND FRIEND, M. T. 1956. The organic matter and nitrogen status of East African soils. *J. Soil Sci.,* **7,** 156-167.
BIROT, Y. AND GALABERT, J. 1967. L'amelioration des rendements en agriculture par amenagements antierosifs et techniques culturales visant a la conservation de l'eau et du sol dans la region de l'Adder-Doutchi Maggia (Rep. de Niger), Station d'Allokoto. Premiers observations, 1966. *Colloque sur la fertilite des sols tropicaux, Tananarive,* **2,** 1316-1331.
BLONDEL, D. 1965a. Premiers elements sur l'influence de la densite apparente du sol sur la croissance racinaire de l'arachide et du sorgho. Ses consequences sur le rendements. *Proc. OAU/STRC Symposium on the maintenance and improvement of soil fertility, Khartoum,* 173-181.
BLONDEL, D. 1965b. Influence du travail du sol sur le profil cultural et les cultures. IRAT/Senegal; rap. an. Div. d'Agropedologie, 427-436 (quoted by Charreau and Nicou, 1971c).
BLONDEL, D. 1966. Premiers resultats sur la dynamique de l'azote dans deux sols du Senegal. Rep. de Recherches 1966. IRAT/Senegal, doc. mimeo, 51 pp. (quoted by Charreau and Nicou, 1971c).
BLONDEL, D. 1967. Premiers resultats sur la dynamique de l'azote mineral de deux sols du Senegal. *Colloque sur la fertilite des sols tropicaux, Tananarive,* **1,** 490-499.
BLONDEL, D. 1970. Relation entre le "nanisme jaune" de l'arachide en sol sableux (dior) et le pH. Definition d'un seuil pour l'activite du *Rhizobium. Agron. trop., Paris,* **26,** 589-595.
BLONDEL, D. 1971a. Contribution a l'etude du lessivage de l'azote en sol sableux (dior) au Senegal. *Agron. trop., Paris,* **26,** 687-696.
BLONDEL, D. 1971b. Contribution a l'etude de la croissance-matiere seche et de l'alimentation azote des cereales de culture seche au Senegal. *Agron. trop., Paris,* **26,** 707-720.
BLONDEL, D. 1971c. Contribution a la connaissance de la dynamique de l'azote mineral en sol sableux (dior) au Senegal. *Agron. trop., Paris,* **26,** 1303-1333.
BLONDEL, D. 1971d. Contribution a la connaissance de la dynamique de l'azote en sol ferrugineux tropical a Sefa (Senegal). *Agron. trop., Paris,* **26,** 1334-1353.
BLONDEL, D. 1971e. Contribution a la connaissance de la dynamique de l'azote mineral en sol ferrugineux tropical a Nioro-du-Rip (Senegal). *Agron. trop., Paris,* **26,** 1354-1361.
BLONDEL, D. 1971f. Role de la plante dans l'orientation de la dynamique de l'azote en sol sableux. *Agron. trop., Paris,* **26,** 1362-1371.
BOCKELEE-MORVAN, A. 1964. Etudes sur la carence potassique de l'arachide au Senegal. *Oleagineux,* **19,** 603-609.
BOCKELEE-MORVAN, A. 1965a. Importance des differentes elements mineraux dans la nutrition minerale de l'arachide. *Proc. 2nd meeting on soil fertility and fertilizer use in W. Africa,* **1,** 150-159.
BOCKELEE-MORVAN, A. 1965b. Efficacite des diverses formes d'apportes des elements mineraux sur l'arachide. *Proc. 2nd meeting on soil fertility and fertilizer use in W. Africa,* **2,** 290-297.
BOCKELEE-MORVAN, A. 1966. Efficacite des diverses formes d'apports des elements mineraux sur l'arachide. *Oleagineux,* **21,** 163-166.
BOCKELEE-MORVAN, A. AND MARTIN, G. 1966a. (Sulphur requirements of the groundnut. Effects on yield). *Oleagineux,* **21,** 679-682.
BOCKELEE-MORVAN, A. AND MARTIN, G. 1966b. Sulphur requirements of peanuts. *Sulphur Inst.* J., **2,** 2-8.
BOCQUIER, G. AND CLAISSE, G. 1963. Reconnaissance pedologique dans les vallees de la Gambie et de la Koulounton (Republique de Senegal). *Cah. ORSTOM,* Ser. pedol., **1,** 4, 5-31.
BOLLE-JONES, E. W. 1964. Incidence of sulphur deficiency in Africa: a review. *Emp. J. expl. Agric.,* **32,** 241-248.

BONFILS, P., CHARREAU, C. AND MARA, M. 1962. Etudes lysimetriques au Senegal. *Agron. trop., Paris*, **17**, 881-914.
BONO, M. 1970. Pennisetum millet and sorghum: synthesis of results. *Afr. Soils*, **15**, 237-248.
BOTHA, A. D. P. 1963. The influence of grass (*Eragrostis curvula*) on the ammonium and nitrate nitrogen content of a black clay soil. *S. Afr. J. agric. Sci.*, **6**, 3-19.
BOUCHET, R. 1963. Les mils et sorghos dans la Republique du Mali. *Agron. trop., Paris*, **18**, 85-107.
BOUCHY, C. 1970. Contribution to the study of soil mineral deficiencies in Ivory Coast cotton growing. *Cot. Fib. trop.*, **25**, fasc. 2 (English edition).
BOUCHY, C. 1973. Essai de fertilisation organo-minerale. Resultats apres dix annees de culture continue mais-cotonnier en Cote d'Ivoire. *Cot. Fib. trop.*, **28**, 343-364.
BOUGHEY, S. A., MUNRO, P. E., MEIKLEJOHN, J., STRONG, R. M. AND SWIFT, M. J. 1964. Antibiotic reactions between African savanna species. *Nature, London*, **203**, 1302.
BOULET, R. 1968. *Etude pedologique de la Haute-Volta region: Centre Nord.* ORSTOM Hann, Dakar.
BOULVERT, Y. 1968. Quelques aspects de l'influence de la topographie et du materiau originel sur la repartition de sols ferrallitiques, sols ferrugineux tropicaux et vertisols dans la region de Bossangoa au nord-ouest de la Republique Centrafricain. *Cah. ORSTOM, Ser. pedol.*, **6**, 259-276.
BOURKE, D. O'D. 1965. Rice in West African sahel and sudan zones. *Afr. Soils*, **10**, 43-55.
BOUYER, S. 1959. Etude de l'evolution du sols dans un secteur de modernisation agricole au Senegal. *Proc. 3rd inter-Afr. Soils Conf.*, **2**, 841-850.
BOUYER, S. 1969. Compte rendu d'une mission effectuee au Dahomey. *Agron. trop., Paris*, **24**, 816-826.
BOUYER, S. AND DAMOUR, M. 1964. Les formes du phosphore dans quelques types des sols tropicaux. *Trans. 8th int. Cong. Soil Sci.*, **4**, 551-561.
BOYER, J. 1972. Soil Potassium, pp. 102-135 in *Soils of the Humid Tropics*. Natl. Acad. Sci., Washington.
BRAMMER, H. 1962. Soils, pp. 88-126 in *Agriculture and Land Use in Ghana*, ed. Wills, J. B. Oxford Univ. Press.
BRAMMER, H. 1967. Soils of the Accra plains. Memoir No. 3, Soil Research Inst., Kumasi, Ghana.
BRAMMER, H. AND ENDREDY, A. S. DE. 1954. The tropical black earths of the Gold Coast and their associated vlei soils. *Trans. 5th int. Cong. Soil Sci.*, **4**, 70-76.
BRAUD, M. 1967. La determination des deficiences minerales dans la nutrition du cotonnier. *Colloque sur la fertilite des sols tropicaux, Tananarive*, **1**, 198-210.
BRAUD, M. 1969-70. Sulphur fertilization of cotton in tropical Africa. *Sulphur Inst. J.*, **5**, 3-5.
BRAUD, M. 1971. Potassium fertilization for cotton in tropical Africa. *Fertilite*, **39**, 5-16.
BRAUD, M., FRITZ, A., MEGIE, C. AND QUILLON, P. J., 1969. Sur la deficience en bore du cotonnier. Observations preliminaires. *Cot. Fib. trop.*, **24**, 465-467.
BREDERO, T. Y. 1966. Fertilizer responses in swamp rice (*Oryza sativa* L.) in Northern Nigeria. *Expl. Agric.*, **2**, 33-43.
BROADBENT, F. E. AND CLARK, F. E. 1965. Denitrification, pp. 344-359 in *Soil Nitrogen*, ed. Bartholomew, W. V. and Clark, F. E. Am. Soc. Agron., Madison. Wisconsin.
BROCKINGTON, N. R. 1961. Studies of the growth of a *Hyparrhenia*-dominant grassland in Northern Rhodesia, II Fertilizer responses, III The effect of fire. *J. br. Grassld. Soc.*, **16**, 54-64.
BROMFIELD, A. R. 1967. End of tour report. Soils and Chem. Section, Res. Div. Ministry of Agric. and Nat. Resources, Western State of Nigeria, Ibadan.
BROMFIELD, A. R. 1972. Sulphur in northern Nigerian soils. 1. The effects of cultivation and fertilizers on total S and sulphate patterns in soil profiles. *J. agric. Sci., Camb.*, **78**, 465-470.
BROMFIELD, A. R. 1973. Uptake of sulphur and other nutrients by groundnuts (*Arachis hypogaea*) in northern Nigeria. *Expl. Agric.*, **9**, 55-58.
BROMFIELD, A. R., 1974a. The deposition of sulphur in dust in northern Nigeria *J. agric. Sci., Camb.*, **83**, 423-425.

BROMFIELD, A. R. 1974b. The deposition of sulphur in the rainwater in northern Nigeria. *Tellus*, **26**, 408-411.

BUNTJER, B. J. 1970. Aspects of the Hausa system of cultivation. *Proc. Seminar on Intercropping, Ford Foundation/IITA/IRAT, Ibadan*, 1970.

CARROLL, D. M. AND KLINKENBERG, K. 1972. Soils, pp. 85-120 in *The Land Resources of North-Eastern Nigeria:* Vol. 1, *The Environment*, ed. Tuley, P. Land Resources Division, Tolworth, England.

CHALK, A. T. 1963. Soil conservation in northern Nigeria and a suggested programme. USAID Rep. C—34.

CHAMINADE, R. 1970. Agricultural research work carried out by IRAT. *Agron. trop., Paris*, **25**, 132-140.

CHANG, S. C. AND JACKSON, M. L. 1957. Fractionation of soil phosphorus. *Soil Sci.*, **84**, 133-144.

CHAROY, J. 1971. Les cultures irrigues au Niger. Resultats de sept annees de mesures et d'experimentations (1963-1970) a la Station Experimentale d'Hydraulique Agricole (Seha) de Tarna dans le Goulbi de Maradi. *Agron. trop., Paris*, **26**, 979-1002.

CHARREAU, C. 1965. Dynamique de l'eau dans deux sols du Senegal. *Agron. trop., Paris*, **16**, 504-561.

CHARREAU, C. 1968. Rain and erosion. *Afr. Soils*, **13**, 241-244.

CHARREAU, C. 1969. Influence des techniques culturales sur le developpement du ruissellement et de l'erosion en Casamance. *Agron. trop., Paris*, **24**, 836-842.

CHARREAU, C. 1972. Problemes poses par l'utilisation agricole des sols tropicaux par des cultures annuelles. *Paper read at Ford Foundation/IRAT/IITA symposium on Tropical Soil Research, Ibadan*, May 1972.

CHARREAU, C. AND FAUCK, F. 1970. Mise au point sur l'utilisation agricole des sols de la region de Sefa (Casamance). *Agron. trop., Paris*, **25**, 151-191.

CHARREAU, C. AND NICOU, R. 1971a,b,c,d. L'amelioration du profil cultural dans les sols sableux et sablo-argileux de la zone tropicale seche ouest-africaine et ses incidences agronomiques (d'apres les travaux des chercheurs d l'IRAT en Afrique de l'Ouest). *Agron. trop., Paris*, **26**, 209-255, 565-631, 903-978, 1183-1237.

CHARREAU, C. AND POULAIN, J. 1963. La fertilisation des mils et sorghos. *Agron. trop., Paris*, **18**, 53-63.

CHARREAU, C. AND POULAIN, J. 1964. Manuring of millet and sorghum. *Afr. Soils*, **9**, 177-191.

CHARREAU, C. AND SEGUY, L. 1969. Mesure de l'erosion et du ruissellement a Sefa en 1968. *Agron. trop., Paris*, **24**, 1055-1097.

CHARREAU, C. AND TOURTE, R. 1967. Le role des facteurs biologiques dans l'amelioration du profil cultural dans les systemes d'agriculture traditionnel de zone tropicale seche. *Colloque sur la fertilite des sols tropicaux, Tananarive*, **2**, 1498-1517.

CHARREAU, C. AND VIDAL, P. 1962. Facteurs pedologique influant sur la croissance et la nutrition des doliques. *Agron. trop., Paris*, **17**, 765-775.

CHARREAU, C. AND VIDAL, P. 1965. Influence de l'*Acacia albida* Del. sur le sol, nutrition minerale et rendements de mils pennisetum au Senegal. *Agron. trop., Paris*, **20**, 600-626.

CHARTER, C. F. 1955. The mechanization of peasant agriculture and the maintenance of soil fertility with bush fallow. Gold Coast Dept. Soil Land-use Surv. Conf. Paper.

CHASE, F. E., CORKE, C. T. AND ROBINSON, J. B. 1968. Nitrifying bacteria in soil, pp. 592-611 in *The Ecology of Soil Bacteria*, ed. Gray, T. R. G. and Parkinson, D. Liverpool Univ. Press.

CHATELIN, Y. 1969. Contribution a l'etude de la sequence sols ferrallitiques rouges et ferrugineux tropicaux beiges. Examen de profiles Centrafricains. *Cah. ORSTOM, Ser. pedol.*, **7**, 447-493.

CHRISTOI, R. 1966. Mesure de l'erosion en Haute-Volta. *Oleagineux*, **21**, 531-534.

CLARK, F. E. 1949. Soil micro-organisms and plant roots. *Adv. Agron.*, **1**, 242-288.

COCHEME, J. AND FRANQUIN, P. 1967a. *A study of the agro-climatology of the semi-arid area south of the Sahara in West Africa.* FAO/UNESCO/WMO Interagency Project on Agroclimatology, Rome.

COCHEME, J. AND FRANQUIN, P. 1967b. *An agroclimatology survey of a semi-arid area in Africa south of the Sahara*. Tech. Note No. 86, World Meteorol. Org. (WMO No. 210, T. P. 110), Geneva.

COMBEAU, A. AND MONNIER, G. 1961. A method for studying structural stability; its application to tropical soils. *Afr. Soils*, **6**, 5-32.

COMBEAU, A. AND QUANTIN, P. 1963. (Observations on seasonal variations of structural stability of tropical soils.) *Cah. ORSTOM, Ser. pedol.*, **1**, 3, 17-26.

COMBEAU, A. AND QUANTIN, P. 1964. Observations sur les relations entre stabilite structurale et materiere organique dans quelques sols d'Afrique Centrale. *Cah. ORSTOM, Ser. pedol.*, **2**, 1, 3-9.

COOKE, G. W. 1967. *The Control of Soil Fertility*. Crosby Lockwood, London.

COOPER, P. J. M. 1971. *Organic Sulphur Fractions in Northern Nigerian Soils*. Ph.D. Thesis, Reading Univ.

CORMACK, R. M. M. 1951. The mechanical protection of arable land. *Rhod. agric. J.*, **48**, 135-164.

COUEY, M., BOUYER, S., CHABROLIN, R. AND COURTESSOLE, P. 1967. Reponse du riz a la fumure dans la region du fleuve Senegal. *Colloque sur la fertilite des sols tropicaux, Tananarive*, **1**, 720-730.

COUEY, M., CHABROLIN, R., BOUYER, S. AND DEJARDIN, J. 1965. Etudes recentes sur la fertilisation du riz dans le delta du Senegal. *Proc. 2nd meeting on soil fertility and fertilizer use in W. Africa*, **2**, 298-312.

C. R. A. BAMBEY, 1957. Phosphate du fond et pluviometrie. Annales du Centre Recherches Agronomiques de Bambey.

CURTIS, D. L. 1965. Sorghum in West Africa. *Field Crop Abstr.*, **18**, 145-152.

CURTIS, D. L. 1966. *Sorghum in Northern Nigeria*. Ph. D. Thesis, Reading Univ.

DABIN, B. 1963. Appreciation des besoins au phosphore dans les sols tropicaux. Les formes du phosphore dans les sols de Cote d'Ivoire. *Cah. ORSTOM, Ser. pedol.*, **1**, 3, 27-42.

DABIN, B. 1967. Sur une methode d'analyse du phosphore dans les sols tropicaux. *Colloque sur la fertilite des sols tropicaux, Tananarive*, **1**, 99-115.

DABIN, B. 1970. Methode d'etude de la fixation du phosphore sur les sols tropicaux. *Cot. Fib. trop.*, **25**, 213-234; 289-310.

DABIN, B. 1971. The use of phosphate fertilizers in a long-term experiment on ferrallitic soil at Bambari, Central African Republic. *Phos. in Agric.*, No. 58, 1-11.

DAGG, M. AND MACARTNEY, J. C. 1968. The agronomic efficiency of the NIAE mechanized tied ridge system of cultivation. *Expl. Agric.*, **4**, 279-294.

DANCETTE, C. AND POULAIN, J. F. 1969. Influence of *Acacia albida* on pedoclimatic factors and crop yields. *Afr. Soils*, **14**, 143-184.

DART, P. J. AND DAY, J. M., 1975. Non-symbiotic nitrogen fixation in the field, in *Soil Microbiology*, ed. Walker, N. Butterworths, London (in press).

DAUBENMIRE, R. 1968. Ecology of fire in grassland. *Adv. Ecol. Res.*, **5**, 209-266.

DAVIES, J. A. 1966. Solar radiation estimates for Nigeria. *Niger. geog. J.*, **9**, 85-100.

DEFFONTAINES, J. P. 1965. Observations sur le profil cultural du sol en conditions diverses. IRAT/Senegal; Doc. Mimeo. 26 pp. (quoted by Charreau and Nicou, 1971c).

DELBOSC, G. 1965. Effet de la fumure organique en zone arachidiere au Senegal. *Proc. 2nd meeting on soil fertility and fertilizer use in W. Africa*, **1**, 99-109.

DELBOSC, G. 1967. Etudes sur la regeneration de la fertilite du sol dans la zone arachidiere du Senegal. *Colloque sur la fertilite des sols tropicaux, Tananarive*, **2**, 1823-1834.

DENNISON, E. B. 1959. The maintenance of soil fertility in the southern Guinea zone of Northern Nigeria (Tiv country). *Trop. Agric., Trin.*, **36**, 171-176.

DENNISON, E. B. 1960. The possible effects of changes in land tenure and soil fertility in Northern Nigeria. *Emp. J. expl. Agric.*, **28**, 94-98.

DENNISON, E. B. 1961. The value of farmyard manure in maintaining fertility in Northern Nigeria. *Emp. J. expl. Agric.*, **29**, 330-336.

D'HOORE, J. L. 1964. *Soil Map of Africa, Scale 1 to 5,000,000*. Joint project no. 11, Commission for Technical Co-operation in Africa, Lagos.

D'HOORE, J. L. 1968. The classification of tropical soils, pp. 7-28 in *The Soil Resources of Tropical Africa*, ed. Moss R. P. Cambridge Univ. Press.

DIAMOND, W. E. DE B. 1937. Fluctuations in nitrogen content of some Nigerian soils. *Emp. J. expl. Agric.*, **5**, 264-280.

DIX, A. N. R. K. 1967. Crop production improvement in the Greater Kano area. FAO Report to the Government of Nigeria.

DIXEY, F. 1963. Geology, applied geology (mineral resources) and geophysics in Africa, pp 51-100 in *A Review of the Natural Resources of the African Continent.* UNESCO, Paris.

DJOKOTO, R. K. AND STEPHENS, D. 1961a. Thirty long-term experiments under continuous cropping in Ghana. I. Crop yields and responses to fertilizers and manures. *Emp. J. expl. Agric.*, **29**, 181-195.

DJOKOTO, R. K. AND STEPHENS, D. 1961b. Thirty long-term fertilizer experiments under continuous cropping in Ghana. II. Soil studies in relation to the effects of fertilizers and manures on crop yields. *Emp. J. expl. Agric.*, **29**, 245-258.

DÖBEREINER, J., DAY, J. M. AND DART, P. J. 1972a. Nitrogenase activity and oxygen sensitivity of the *Paspalum notatum-Azotobacter paspali* association. *J. gen. Microbiol.*, **71**, 103-116.

DÖBEREINER, J., DAY, J. M. AND DART, P. J. 1972b. Nitrogenase activity in the rhizosphere of sugar cane and some other tropical grasses. *Pl. Soil.*, **37**, 191-196.

DOMMERGUES, Y. 1954. (Effect of fire on the microflora of grassland soils). *Mem. Inst. sci. Madagascar Ser. D6*, 149-158.

DOMMERGUES, Y. 1956a. Action d'amendements calciques et phosphates sur l'activite biologique de deux sols du Senegal. *Trans. 6th int. Cong. Soil Sci.*, **C**, 381-387.

DOMMERGUES, T. 1956b. Etude de la biologie des sols des forets tropicales seches et de leur evolution apres defrichement. *Trans. 6th int. Cong. Soil Sci.*, **E**, 605-610.

DOMMERGUES, Y. 1959. Characteristiques biologiques de quelques grands types de sol de l'ouest Africain. *Proc. 3rd inter-Afr. Soils Conf.*, **1**, 215-220.

DOMMERGUES, Y. 1960. Un exemple d'utilisation des techniques biologiques dans le characterisation des types pedologiques. *Agron. trop., Paris*, **15**, 61-72.

DOMMERGUES, Y. 1966. (Symbiotic fixation of nitrogen by *Casuarina*). *Suppl. Annls. Inst. Pasteur, Paris*, **111**, no. 3, 247-258 (*Soils Fertil.*, **30**, 344).

DOMMERGUES, Y., BALANDREAU, J., RINAUDO, G. AND WEINHARD, P. 1973. Non-symbiotic nitrogen fixation in the rhizosphere of rice, maize and different tropical grasses. *Soil Biol. Biochem.*, **5**, 83-89.

DOYNE, H. C., HARTLEY, K. T. AND WATSON, W. A. 1938. Soil types and manurial experiments in Nigeria. *Proc. 3rd West Afr. agric. Conf.*, **1**, 227-298.

DUMONT, C. 1966. Les recherches rizicoles en Haute-Volta de 1959 a 1965. *Agron. trop., Paris*, **21**, 558-565.

DUMONT, C. AND DUPONT DE DINECHIN, B. 1967. La fumure phosphatee des cultures vivrieres en Haute-Volta. *Colloque sur la fertilite des sols tropicaux, Tananarive*, **1**, 639-656.

DUPONT DE DINECHIN, B. 1967a. Contribution a l'etude des exportations du mais et du sorgho en Haute Volta. *Colloque sur la fertilite des sols tropicaux, Tananarive*, **1**, 528-543.

DUPONT DE DINECHIN, B. 1967b. Observations sur l'interet des phosphates naturels pour la fumure de cereales en Haute Volta. *Colloque sur la fertilite des sols tropicaux, Tananarive*, **1**, 657-668.

DUPONT DE DINECHIN, B. 1967c. La fumure potassique des cultures vivrieres en Haute Volta. *Colloque sur la fertilite des sols tropicaux, Tananarive*, **1**, 669-677.

DUPONT DE DINECHIN, B. 1967d. Resultats concernant les effets compares des fumures minerale et organique. *Colloque sur la fertilite des sols tropicaux, Tananarive*, **2**, 1411-1428.

DURAND, J. H. 1965. Etude pedologique de la plaine de Boghe (Republique Islamique de Mauritanie). *Agron. trop., Paris*, **20**, 1284-1316.

EDWARDS, D. C. 1942. Grass-burning. *Emp. J. expl. Agric.*, **10**, 219-231.

ENDREDY, A. S. DE. 1954. The organic matter content of Gold Coast soils. *Trans. 5th int. Cong. Soil Sci.*, **2**, 457-463.

ENWEZOR, W. O. 1967. Soil drying and organic matter decomposition. *Pl. Soil*, **26**, 269-276.

ENWEZOR, W. O. AND MOORE, A. W. 1966a. Comparison of two methods for determining organic phosphorus in some Nigerian soils. *Soil Sci.*, **102**, 284-285.

ENWEZOR, W. O. AND MOORE, A. W. 1966b. Phosphorus status of some Nigerian soils. *Soil Sci.*, **102**, 322-328.
EVANS, A. C. 1963. Soil fertility studies in Tanganyika. III. *E. Afr. agric. For. J.*, **28**, 231-239.
EVELYN. S. H. AND THORNTON, I. 1964. Soil fertility and the response of groundnuts to fertilizers in the Gambia. *Emp. J. expl. Agric.*, **32**, 153-160.

FAO, 1965. Agricultural development in Nigeria, 1965-80. FAO, Rome.
FAO, 1967. Ghana. Land and water survey in the Upper and Northern Regions. Final report. Vol. III.
FAO, 1968. Freedom from Hunger Fertilizer Programme. Rep. L.A.; FFHC/68/14, Rome.
FAO, 1971. *Production Yearbook*, vol. 25. FAO, Rome.
FAO/ICA, 1960. Report on the agricultural survey of Northern Region of Nigeria. Kaduna.
FAUCK, R. 1956. Evolution des sols sous culture mecanisee dans les regions tropicales. *Trans. 6th int. Cong. Soil Sci.*, E, 593-596.
FAUCK, R. 1960. Matiere organique et azote des sols de la Moyenne-Guinea et relations avec les rendements des cultures. *C. R. Acad. Agric. Fr.*, **46**, 152-155.
FAUCK, R. 1963. The sub-group of leached ferruginous tropical soils with concretions. *Afr. Soils*, **8**, 407-429.
FAUCK, R. 1964. Les sols rouge faiblement ferrallitiques d'Afrique Occidentale. *Trans. 8th int. Cong. Soil Sci.*, **5**, 569-574.
FAUCK, R., MOUREAUX, C. AND THOMANN, C. 1969. Bilans de l'evolution des sols de Sefa (Casamance, Senegal) apres quinze annees de culture continue. *Agron. trop., Paris*, **24**, 263-301.
FAULKNER, R. C. 1971. Progress Reports, Season 1969-70, Northern States, Nigeria. *Cotton Grow. Rev.*, **48**, 72.
FAULKNER, R. C., SMITHSON, J. B. AND LANGHAM, R. M. 1970. Cotton Research Corporation, Progress Report, Northern States of Nigeria, 1969-70, p. 21.
FERGUSON, H. 1944. Deterioration of soils and vegetation in Equatoria Province with special reference to grass fires. Sudan Govt. Soil Conserv. Cttee. Rpt. 1944 pp. 138-142 (*Soils Fertil.*, **9**, 102).
FLOYD, B. 1965. Soil erosion and deterioration in Eastern Nigeria. *Niger. geogr. J.*, **8**, 33-44.
FOURNIER, F. 1960. *Climat et Erosion. La relation entre l'erosion du sol par l'eau et les precipitations atmospheriques.* These Doct. es-lettres, PUF, Paris.
FOURNIER, F. 1962. Carte du danger d'erosion en Afrique au sud du Sahara. CEE-CCTA.
FOURNIER, F. 1963. The Soils of Africa, pp. 221-248 in *A Review of the Natural Resources of the African Continent.* UNESCO, Paris.
FOURNIER, F. 1967. Research on soil erosion and soil conservation in Africa. *Afr. Soils*, **12**, 53-96.
FOURRIER, P. 1965. Les oligo-elements dans la culture de l'arachide du Nord-Senegal. *Proc. 2nd meeting on soil fertility and fertilizer use in W. Africa*, **2**, 352-359.
FOX, R. L., PLUCKNETT, D. L. AND WHITNEY, A. S. 1968. Phosphate requirements of Hawaiian latosols and residual effects of fertilizer phosphorus. *Trans. 9th int. Cong. Soil Sci.*, **2**, 301-310.
FRANKEL, F. R. 1972. *India's Green Revolution: Economic Gains and Political Costs.* Princeton Univ. Press.
FRENEY, J. R. 1967. Sulphur containing organisms, pp. 229-259 in *Soil Biochemistry*, ed. McLaren, A. D. and Peterson, G. H. Dekker, New York.
FRIED, M. AND BROESHART, H. 1967. *The Soil-Plant System in Relation to Inorganic Nutrition.* Acad. Press, New York.
FRIEND, M. T., AND BIRCH, H. F. 1960. Phosphate responses in relation to soil tests and organic phosphorus. *J. agric. Sci., Camb.*, **54**, 341-347.
FRITZ, A. 1971. La deficience en bore du cotonnier au Nord-Cameroun. *Cot. Fib. trop.*, **25**, 235-241.
FURON, R. 1963. *Geology of Africa* (trans. Hallam, H. and Stevens, L.A.). Oliver and Boyd, London.

Gadet, R. 1967. Mineralisation et nitrification de l'azote dans quelques sols tropicaux. *Colloque sur la fertilite des sols tropicaux, Tananarive*, **1**, 505-507.

Galland, P. 1963. Vulgarisation apres des cultivateurs Voltaiques des resultats de la recherche agronomique sur l'arachide. *Oleagineux*, **18**, 771-775.

Ganry, F. 1972. The date problem for the sowing of millet (*Pennisetum typhoides*) in the tropical zone. *Afr. Soils*, **17**, 53-57.

Gautreau, J. 1965. Influence du regime des eaux sur l'efficacite des engrais. *Proc. 2nd meeting on soil fertility and fertilizer use in W. Africa*, **1**, 142-149.

Gavaud, M. 1968. Les sols bien draines sur materiaux sableux du Niger. *Cah. ORSTOM, Ser. pedol.*, **6**, 277-307.

Geus, J. G. de 1967. *Fertilizer guide for tropical and subtropical farming*. Centre d'Etude de l'Azote, Zurich.

Gillier, P. 1964. Les coques d'arachides; leur utilisation en agriculture. *Oleagineux*, **19**, 473-475.

Gillier, P. 1966. L'arachide et le molybdene. *C.r. Seanc. hebd. Acad. Agric., Fr.*, **52**, 446-448.

Gillier, P. and Gautreau, H. 1971. Dix ans d'experimentation dans la zone a carence potassique de Patar au Senegal. *Oleagineux*, **26**, 33-38.

Goldschmidt, V. 1962. *Geochemistry*, ed. Muir, A. Clarendon Press ,Oxford.

Goldson, J. R. 1963. The effect of time of planting on maize yields. *E. Afr. agric. For. J.*, **29**, 160-163.

Goldsworthy, P. R. 1967a. Responses of cereals to fertilizers in northern Nigeria. I. Sorghum. *Expl. Agric.*, **3**, 29-40.

Goldsworthy, P. R. 1967b. Responses of cereals to fertilizer in northern Nigeria. II. Maize. *Expl. Agric.*, **3**, 263-273.

Goldsworthy, P. R. 1967c. Unpublished report at Cropping Scheme Meeting (1967), Inst. Agric. Research, Samaru, Nigeria.

Goldsworthy, P. R. 1968. Interim results of a long-term phosphate residual experiment. Unpublished report at Cropping Scheme Meeting (1968), Inst. Agric. Research, Samaru, Nigeria.

Goldsworthy, P. R. and Heathcote, R. G. 1963. Fertilizer trials with groundnuts in northern Nigeria. *Emp. J. expl. Agric.*, **31**, 351-366.

Goldsworthy, P. R. and Heathcote, R. G. 1964. Fertilizer trials with soya beans in northern Nigeria. *Emp. J. expl. Agric.*, **32**, 257-262.

Goldsworthy, P. R. and Meredith, R. M. 1959. Some investigations on the maintenance of soil fertility—Northern Nigeria. Duplicated paper for Agric. Officers' Conference, Oct. 1959.

Greenland, D. J. 1956. Is aerobic denitrification important in tropical soils? *Trans. 6th int. Cong. Soil Sci.*, B, 765-769.

Greenland, D. J. 1958. Nitrate fluctuations in tropical soils. *J. agric. Sci., Camb.*, **50**, 82-92.

Greenland, D. J. 1962. Denitrification in some tropical soils. *J. agric. Sci., Camb.*, **58**, 227-233.

Greenland, D. J. 1965. Interaction between clays and organic compounds in soils. Part II. Adsorption of soil organic compounds and its effect on soil properties. *Soils Fertil.*, **28**, 521-532.

Greenland, D. J. 1971. Changes in the nitrogen status and physical condition of soils under pastures, with special reference to the maintenance of the fertility of Australian soils used for growing wheat. *Soils Fertil.*, **34**, 237-251.

Greenland, D. J. 1972. Soil factors determining responses to phosphorus and nitrogen fertilizers used in tropical Africa. *Afr. Soils*, **17**, 99-108.

Greenland, D. J. and Nye, P. H. 1959. Increases in the carbon and nitrogen contents of tropical soils under natural fallows. *J. Soil Sci.*, **10**, 284-299.

Greenland, D. J. and Nye, P. H. 1960. Does straw induce nitrogen deficiency in tropical soils? *Trans. 7th int. Cong. Soil Sci.*, **2**, 478-485.

Greenwood, M. 1948. Mixed farming and fertilizers in Northern Nigeria. *Proc. trop. Soil Conf., Harpenden, England*, 170-173.

Greenwood, M. 1951. Fertilizer trials with groundnuts in Northern Nigeria. *Emp. J. expl. Agric.*, **19**, 225-241.

Greenwood, M. 1954. Sulphur deficiency in groundnuts in Northern Nigeria. *Trans. 5th int. Cong. Soil Sci.*, **3**, 245-251.

GRIFFITH, G. AP 1951. Factors influencing nitrate accumulation in Uganda soil. *Emp. J. expl. Agric.*, **19**, 1-12.
GRIFFITHS, J. F. (ed.) 1971. *Climates of Africa*, vol. 10 of *World Survey of Climatology*, ed.-in-chief Landsberg, H. E. Elsevier, Amsterdam.
GUICHARD, E., BOUTEYRE, G. AND LEPOUTRE, B. 1959. Etude pedologique des polders de Bol et Bolguini. ORSTOM mimeo.
GUILLOTEAU, J. 1954. A problem in rural economy: the improvement of agricultural economy in Senegal. *Afr. Soils*, **3**, 12-39.
GUILLOTEAU, J. 1956. The problem of bush fires and burning in land development and soil conservation in Africa south of the Sahara. *Afr. Soils*, **4**, 64-102.
GUISCAFRE, J. 1961. Conservation des sols et protection des cultures par bandes brise-vent. Cantons Doukoula, Tchatibali et Wina (Cameroun). *Bois For. trop.*, **79**, 17-29 (*Fld. Crop Abstr.*, **15**, 2108).

HAAS, H. J., EVANS, C. E. AND MILES, E. F. 1957. Nitrogen and carbon changes in Great Plains soils as influenced by cropping and soil treatments. Tech. Bull. 1164, U. S. Dept. Agric., Washington.
HADDAD, G. AND SEGUY, L. 1972. Le riz pluvial dans le Senegal meridional. Bilan de quatre annees d'experimentation (1966-1969). *Agron. trop., Paris*, **27**, 419-461.
HAGENZIEKER, F. 1957. Soil nitrogen studies at Urambo, Tanganyika Territory, East Africa. *Pl. Soil*, **9**, 97-113.
HALM, A. T. 1968. Correlation of soil tests for available phosphorus with crop yield. *Ghana J. agric. Sci.*, **1**, 29-33.
HAMILTON, R. 1966. Aeolian soils of the West African Sudan. *Trans. Conf. Medit. Soils, Madrid*, 21-23.
HARKNESS, C. 1970. The trade in Nigerian groundnuts. *Samaru Agric. Newsletter*, **12**, 62-65. Inst. Agric. Res., Samaru, Nigeria.
HARMSEN, G. W. AND KOLENBRANDER, G. J. 1965. Soil inorganic nitrogen, pp. 43-92, in *Soil Nitrogen*, ed. Bartholomew, W. V. and Clark, F. E. Am. Soc. Agron., Madison, Wisconsin.
HARRADINE, F. AND JENNY, H. 1958. Influence of parent material and climate on texture and nitrogen and carbon contents of virgin California soils. 1. Texture and nitrogen contents of soils. *Soil Sci.*, **85**, 235-243.
HARTLEY, A. C. 1961. Nitrogen transformations in a Trinidad clay soil. D. T. A. Dissertation, Univ. of West Indies, Trinidad (quoted by Odu & Vine, 1968).
HARTLEY, K. T. 1937. An explanation of the effect of farmyard manure in Northern Nigeria. *Emp. J. expl. Agric.*, **5**, 254-263.
HASLE, H. 1965. Les cultures vivrieres au Dahomey. *Agron. trop., Paris*, **20**, 725-746.
HAUGHTON, S. M. 1963. *The Stratigraphical History of Africa South of the Sahara.* Oliver and Boyd, Edinburgh and London.
HAWKINS, P. AND BRUNT, M. 1965. *The Soils and Ecology of West Cameroun.* Vol. I. FAO.
HEATHCOTE, R. G. 1970. Soil fertility under continuous cultivation in northern Nigeria. I. The role of organic manures. *Expl. Agric.*, **6**, 229-237.
HEATHCOTE, R. G. 1972a. The effect of potassium and trace elements on yield in northern Nigeria. *Afr. Soils*, **17**, 85-89.
HEATHCOTE, R. G. 1972b. Potassium fertilization in the savanna zone of Nigeria. *Potash Review*, May 1972.
HEATHCOTE, R. G. AND SMITHSON, J. B. 1974. Boron deficiency in cotton in northern Nigeria. I. Factors influencing occurrence and methods of correction. *Expl. Agric.*, **10**, 199-208.
HEATHCOTE, R. G. AND STOCKINGER, K. R. 1970. Soil fertility under continuous cultivation in northern Nigeria. II. Responses to fertilizers in the absence of organic manures. *Expl. Agric.*, **6**, 345-350.
HENIN, S., GRAS, R. AND MONNIER, G. 1969. *Le Profil Cultural.* Masson et cie., Paris.
HESSE, P. R. 1956. Studies of sulphur metabolism in soils and muds. E. Afr. agric. Res. Org. Rep., 1956, pp. 22-28 (*Soils Fertil.*, **21**, 1606).
HIGGINS, G. M. 1963. Upland soils of the Samaru and Kano areas. Samaru Soil Survey Bull. (unnumbered). Inst. Agric. Res., Samaru, Nigeria.
HIGGINS, G. M. 1965a. Pedological factors in northern Nigerian soils. I. The factors. *Proc. 2nd meeting on soil fertility and fertilizer use in W. Africa*, **1**, 29-34.

HIGGINS, G. M. 1965b. Pedological factors in northern Nigerian soils. II. The influence of pedological factors on soil groups. *Proc. 2nd meeting on soil fertility and fertilizer use in W. Africa*, **1**, 35-46.
HIGGINS, G. M. AND OKUNOLA, Y. A. 1965. Report on the detailed survey of the Ochanja area, Kabba Province, Northern Nigeria. Samaru Soil Survey Bull. No. 30 Inst. Agric. Res., Samaru, Nigeria.
HILL, P. R. 1961. A simple system for soil and water conservation and drainage in mechanized crop production in the tropics. *Emp. J. expl. Agric.*, **29**, 337-349.
HINGSTON, F. J., ATKINSON, R. J., POSNER, A. M. AND QUIRK, J. P. 1968. Specific adsorption of anions on goethite. *Trans. 9th int. Cong. Soil Sci.*, **1**, 669-678.
HOMER, D. R. 1962. Ann. Rpt. Min. Agric. Northern Nigeria, 1961-2, Part II.
HOPKINS, B. 1965a. *Forest and Savanna*. Heinemann, Ibadan and London.
HOPKINS, B. 1965b. Observations on savanna burning in the Olokemeji Forest Reserve, Nigeria. *J. appl. Ecol.*, **2**, 367-382.
HOPKINS, B. 1966. Vegetation of the Olokemeji Forest Reserve, Nigeria. IV. The litter and soil with special reference to their seasonal changes. *J. Ecol.*, **54**, 687-703.
HUDSON, N. W. 1957. Erosion control research. *Rhod. agric. J.*, **54**, 297-323.
HUDSON, N. W. 1971. *Soil Conservation*. Batsford, London.
HUDSON, N. W. AND JACKSON, D. C. 1959. Results achieved in the measurement of erosion and runoff in Southern Rhodesia. *Proc. 3rd inter-Afr. Soils Conference, Dalaba.*
HUFFMAN, E. D. 1962. Reactions of phosphate in soils: recent research by T. V. A. Proc. Fertil. Soc., Lond., no. 71.

INST. AGR. RES., SAMARU 1968. Ann. Rpt. Inst. Agr. Res., Samaru, 1967-8. Ahmadu Bello Univ., Zaria, Nigeria.
IPINMIDUN, W. B. 1972. Organic phosphorus in some northern Nigerian soils in relation to soil organic carbon and as influenced by parent rock and vegetation. *J. Sci. Fd. Agric.*, **23**, 1099-1105.
IRAT/HAUTE VOLTA 1972. Principaux resultats agronomiques obtenus par l'Institut de Recherches Agronomique et des Cultures Vivrieres en Haute-Volta, 1960-71, *Proc. Ford Foundation/IITA/IRAT Seminar on Tropical Soil Research, Ibadan.* May 1972.
IRCT 1965a. La fertilisation minerale du cotonnier au Dahomey (Zone Nord). *Proc. 2nd meeting on soil fertility and fertilizer use in W. Africa*, **2**, 336-340.
IRCT 1965b. La fertilisation minerale du cotonnier en culture seche au Mali. *Proc. 2nd meeting on soil fertility and fertilizer use in W. Africa*, **2**, 341-345.
IRCT 1965c. La fertilisation minerale de la culture cotonnier en Cote d'Ivoire. *Proc. 2nd meeting on soil fertility and fertilizer use in W. Africa*, **3**, 349-351.
IRVINE, F. R. 1969. *West African Crops*. Oxford Univ. Press.

JACQUINOT, L. 1964. Contribution a l'etude de la nutrition minerale du sorgho congossane (*sorghum vulgare*, Var. *guineense*). *Agron. trop., Paris*, **19**, 669-722.
JACQUINOT, L. 1965. Fixation des phosphates par les sols de culture de decrue dans la region de Kaedi. Etude au moyen de ^{32}P. *Agron. trop., Paris*, **20**, 593-599.
JAIYEBO, E. O. 1967. Occurrence of non-exchangeable ammonium in soils. *Niger. agric. J.*, **4**, 65-68.
JENNY, H. AND RAYCHAUDHURI, S. P. 1960. Effect of climate and cultivation on nitrogen and organic matter reserves in Indian soils. Ind. Counc. agric. Res., New Delhi.
JENNY, J. J. 1965. Sols et problemes de fertilite en Haute-Volta. *Agron. trop., Paris*, **20**, 220-247.
JENSEN, H. L. 1965. Non-symbiotic nitrogen fixation, pp. 436-480 in *Soil Nitrogen*, ed. Bartholomew, W. V. and Clark, F. E. Am. Soc. Agron., Madison, Wisconsin.
JEWITT, T. N. 1949. Soil fertility studies in the Zande district of the Anglo-Egyptian Sudan. *Commonwealth Bur. Soils Tech. Comm.*, **46**, 165-169.
JOHNSTON, B. F. 1958. *The Staple Food Economies of West Tropical Africa*. Stanford Univ. Press.
JONES, E. 1960. Contribution of rainwater to the nutrient economy of soil in northern Nigeria. *Nature, London*, **188**, 432.
JONES, E. 1968. Nutrient cycle and soil fertility on red ferrallitic soils. *Trans. 9th int. Cong. Soil Sci.*, **3**, 419-427.

JONES, M. J. 1971. The maintenance of soil organic matter under continuous cultivation at Samaru, Nigeria. *J. agric. Sci., Camb.*, **77**, 473-482.
JONES, M. J. 1973. The organic matter content of the savanna soils of West Africa. *J. Soil Sci.*, **24**, 42-53.
JONES, M. J. 1974a. Effects of previous crop on yield and nitrogen response of maize at Samaru, Nigeria. *Expl. Agric.*, **10**, 273-279.
JONES, M. J. 1974b. Observations on dry-season moisture profiles at Samaru. *Samaru misc. Pap. No.* 51.
JONES, M. J. 1975a. Leaching of nitrate under maize at Samaru, Nigeria. *Trop. Agric., Trin.*, **52**, 1-10.
JONES, M. J. 1975b. New recommendations for nitrogen fertilizers in the northern States. *Samaru agric. Newsletter* (in press). Inst. Agric. Res., Samaru, Nigeria.
JONES, M. J. AND BROMFIELD, A. R. 1970. Nitrogen in the rainfall at Samaru, Nigeria. *Nature, London*, **227**, 86.
JONES, M. J. AND STOCKINGER, K. R. 1972. The effect of planting date on the growth and yield of maize at Samaru, Nigeria. *Afr. Soils*, **17**, 27-34.
JORDAN, H. D. 1964. Upland rice—its culture and significance in Sierra Leone. *Int. Rice Comm. Newsl.*, **13**, No. 2, 26-9 (*Field Crop Abstr.*, **18**, 721.).
JUO, A. S. R. AND ELLIS, B. G. 1968. Particle size distribution of aluminium, iron and calcium phosphates in soil profiles. *Soil Sci.*, **106**, 374-380.
JURION, F. AND HENRY, J. 1969. *Can Primitive Farming Be Modernised*? trans. Agr. Europe (London), Publ. l'Inst. Natl. l'Etude Agron. Congo.

KADEBA, O. 1970. Organic matter and nitrogen status of some soils from the savanna zone of Nigeria. *Proc. inaugural Conf. Forestry Assoc., Nigeria. Ibadan*, 1970.
KALOGA, B. 1966. (Pedological study of the slopes of the White and Red Volta basins in Upper Volta). *Cah. ORSTOM, Ser. pedol.*, **4**, 23-61.
KALOGA, B. 1970. Etude pedologique des bassins versants des Voltas Blanche et Rouge (3e partie). *Cah. ORSTOM, Ser. pedol.*, **8**, 187-218.
KASS, D. L., DROSDOFF, M. AND ALEXANDER, M. 1971. Nitrogen fixation by *Azotobacter paspali* in association with bahiagrass (*Paspalum notatum*). *Proc. Soil Sci. Soc. Am.*, **35**, 286-289.
KASSAM, A. H. AND KOWAL, J. M. 1973. Productivity of crops in the savanna and rain forest zone in Nigeria. *Savanna*, **2**, 39-49.
KEAY, R. W. J. 1959. *An Outline of Nigerian Vegetation*. Nigerian Govt. Printer, Lagos.
KEAY, R. W. J. 1960. An example of Northern Guinea zone vegetation in Nigeria. Nigerian For. Inform. Bull. (New Series) No. 1.
KENDREW, W. G. 1961. *The Climates of the Continents*. Oxford Univ. Press.
KING, L. C. 1967. *The Morphology of the Earth: A Study and Synthesis of World Scenery*, 2nd ed. Oliver and Boyd, London.
KINJO, T. AND PRATT, P. F. 1971. Nitrate adsorption: I. In some acid soils of Mexico and South America. *Proc. Soil Sci. Soc. Am.*, **35**, 722-725.
KLEMMEDSON, J. O. AND JENNY, H. 1966. Nitrogen availability in Californian soils in relation to precipitation and parent material. *Soil Sci.*, **102**, 215-222.
KLINKENBERG, K. 1965. Report on the reconnaissance soil survey of part of Borgu Division, Ilorin Province. Samaru Soil Survey Bull. No. 28. Inst. Agric. Res., Samaru, Nigeria.
KLINKENBERG, K. AND HIGGINS, G. M. 1968. An outline of northern Nigerian soils. *Niger. J. Sci.*, **2**, 91-115.
KLINKENBERG, K. AND HILDEBRAND, F. H. 1965. Report on the semi-detailed soil survey of areas near the new town-site of Bussa. Samaru Soil Survey Bull. No. 37. Inst. Agric. Res., Samaru, Nigeria.
KOECHLIN, J. 1963. Flora of Africa south of the Sahara, pp. 263-275 in *A Review of the Natural Resources of the African Continent*. UNESCO, Paris.
KONONOVA, M. M. 1966. *Soil Organic Matter*, 2nd ed. Pergamon, London.
KOWAL, J. 1968a. Physical properties of soil at Samaru, Zaria, Nigeria: storage of water and its use by crops. I. Physical status of soils. *Niger. agric. J.*, **5**, 13-20.
KOWAL, J. 1968b. Physical properties of soils at Samaru, Zaria, Nigeria: storage of water and its use by crops. II. Water storage characteristics. *Niger. agric. J.*, **5**, 49-52.

KOWAL, J. 1969. Some physical properties of soils at Samaru, Zaria, Nigeria: storage of water and its use by crops. III. Seasonal pattern in soil water changes under various vegetation covers and bare fallow. *Niger. agric. J.*, **6**, 18-29.

KOWAL, J. 1970a. The hydrology of a small catchment basin at Samaru, Nigeria. II. Assessment of the main components of the water budget. *Niger. agric. J.*, **7**, 41-52.

KOWAL, J. 1970b. The hydrology of a small catchment basin at Samaru, Nigeria. III. Assessment of surface runoff under varied land management and vegetation cover. *Niger. agric. J.*, **7**, 120-133.

KOWAL, J. 1970c. The hydrology of a small catchment basin at Samaru, Nigeria. IV. Assessment of soil erosion under varied land management and vegetation cover. *Niger. agric. J.*, **7**, 134-147.

KOWAL, J. 1970d. Effect of an exceptional storm on soil conservation at Samaru, Nigeria. *Niger. geog. J.*, **13**, 163-173.

KOWAL, J. AND ANDREWS, D. J. 1973. Pattern of water availability and water requirement for grain sorghum production at Samaru, Nigeria. *Trop. Agric., Trin.*, **50**, 89-100.

KOWAL, J. AND FAULKNER, R. C. 1975. Cotton production in the northern states of Nigeria in relation to water availability and crop water use. *Cotton Grow. Rev.*, **52**, 11-29.

KOWAL, J. AND KNABE, D. T. 1972. *An Agroclimatological Atlas of the Northern States of Nigeria, with Explanatory Notes*. Ahmadu Bello Univ. Press, Zaria, Nigeria.

KOWAL, J. AND OMOLOKUN, A. O. 1970. The hydrology of a small catchment basin at Samaru, Nigeria. I. Seasonal fluctuations in the height of the ground water table. *Niger. agric. J.*, **7**, 27-40.

LAL, R. 1973. Effects of seed bed preparation and time of planting on maize (*Zea mays*) in western Nigeria. *Expl. Agric.*, **9**, 303-313.

LAMOUROUX, M. 1969. *Carte pedologique du Togo*. Notice explicative No. 34. Centre ORSTOM de Lome, Togo.

LARSEN, S. 1967. Soil phosphorus. *Adv. Agron.*, **19**, 151-210.

LARSEN, S., GUNARY, D. AND SUTTON, C. D. 1965. The rate of immobilization of applied phosphate in relation to soil properties. *J. Soil Sci.*, **16**, 141-148.

LAWES, D. A. 1957. Preliminary report on soil-water-crop relationships at Samaru, Northern Nigeria. *Emp. Cotton Grow. Rev.*, **34**, 155-160.

LAWES, D. A. 1961. Rainfall conservation and the yield of cotton in Northern Nigeria. *Emp. J. expl. Agric.*, **29**, 307-318.

LAWES, D. A. 1963. A new cultivation technique in tropical Africa. *Nature, London*, **198**, 1328.

LAWES, D. A. 1964. *Rainfall runoff and soil surface infiltration rates at Samaru, Northern Nigeria*. M. Sc. Thesis, Univ. of Wales.

LAWES, D. A. 1966. Rainfall conservation and the yields of sorghum and groundnuts in Northern Nigeria. *Expl. Agric.*, **2**, 139-146.

LE BUANEC, B. 1972. Influence du labour sur les rendements de riz et de mais en sols ferrallitiques. *Afr. Soils*, **17**, 173-179.

LEEUW, P. N. DE 1966. The role of savanna in nomadic pastoralism: some observations from western Bornu, Nigeria. *Neth. J. agric. Sci.*, **13**, 178-189.

LEMAITRE, C. 1954. Les problemes de la conservation des sols au Niger et le "Gao". *Proc. 2nd inter-Afr. Soils Conf.*, **1**, 569-580.

LE MARE, P. H. 1968. Experiments on the effects of phosphate applied to a Buganda soil. II. Field experiments on the response curve to triple superphosphate. *J. agric. Sci., Camb.*, **70**, 271-279.

LENEUF, N. 1954a. Les sols du secteur cotonniere de Haute-Volta. *Proc. 2nd inter-Afr. Soils Conf.*, **2**, 971-991.

LENEUF, N. 1954b. Les terres noires du Togo. *Trans. 5th int. Cong. Soil Sci.*, **4**, 131-136.

LEVEQUE, A. 1969. Les principaux evenements geomorphologiques et les sols sur le socle granito-gneissique du Togo. *Cah. ORSTOM, Ser. pedol.*, **7**, 203-224.

LIENART, J. M. AND NABOS, J. 1967. Les recherches sur la fertilisation de l'arachide et l'amelioration fonciere des sols au Niger. *Colloque sur la fertilite des sols tropicaux, Tananarive*, **1**, 1046-1057.

MACARTNEY, J. C., NORTHWOOD, P. J., DAGG, M. AND LAWSON, R. 1971. The effect of different cultivation techniques on soil moisture conservation and the establishment and yield of maize at Kongwa, Central Tanzania. *Trop. Agric., Trin.*, **48**, 9-23.

MCEWAN, J. 1962. Raising cereal yields in northern Ghana. *Emp. J. expl. Agric.*, **30**, 330-334.

MCGARITY, J. W. 1959. Some factors affecting the nitrogen economy of a humid sub-tropical soil and a semi-arid temperate soil. *J. Aust. Inst. agric. Sci.*, **25**, 322-323.

MCGARITY, J. W. 1962. Effect of freezing of soil on denitrification. *Nature, London*, **196**, 1342-1343.

MAIGNIEN, R. 1948. La matiere organique et l'eau dans les sols des regions Nord-Ouest du Senegal. *Proc. 1st inter-Afr. Soils Conf.*, **1**, 247-251.

MAIGNIEN, R. 1959. Les sols subarides au Senegal. *Agron. trop., Paris*, **14**, 535-571.

MAIGNIEN, R. 1961. The transition from ferruginous tropical soils to ferrallitic soils in the south-west of Senegal. *Afr. Soils*, **6**, 173-228.

MAIGNIEN, R. 1963. Eutrophic brown soils. *Afr. Soils*, **8**, 491-496.

MARTIN, D., SIEFFERMANN, G. AND VALLERIE, M. 1966. Les sols rouges du Nord-Cameroun. *Cah. ORSTOM, Ser. pedol.*, **4**, 3-28.

MARTIN, G. 1963. Degradation de la structure des sols sous culture mecanises dans la vallee du Niari. *Cah. ORSTOM, Ser. pedol.*, **1**, 2, 8-14.

MARTIN, G. 1970. Synthese agropedologique des etudes ORSTOM dans la vallee du Niari, en Republique du Congo-Brazzaville. *Cah. ORSTOM, Ser. pedol.*, **8**, 63-79.

MARTIN, G. AND FOURRIER, P. 1965. Les oligo-elements dans la culture de l'arachide du Nord Senegal. *Oleagineux*, **20**, 287-291.

MARTIN, W. S. 1944. Soil structure. *E. Afr. agric. J.*, **9**, 189-195.

MEGIE, C., LOUIS, P. AND GUIBERT, P. 1970. Contribution to the study of a mineral fertilizer formula in the soils of Chad. *Cot. Fib. trop.*, **25**, fasc. 2 (English edition).

MEIKLEJOHN, J. 1954. Notes on nitrogen-fixing bacteria from East African soils. *Trans. 5th int. Cong. Soil Sci.*, **3**, 123-125.

MEIKLEJOHN, J. 1955. The effect of bush burning on the microflora of a Kenya upland soil. *J. Soil Sci.*, **6**, 111-118.

MEIKLEJOHN, J. 1962. Microbiology of the nitrogen cycle in some Ghana soils. *Emp. J. expl. Agric.*, **30**, 115-126.

MEIKLEJOHN, J. 1967. Microbiological observations on an experiment with compost. *Rhod. Zamb. Mal. J. agric. Res.*, **5**, 43-55.

MEIKLEJOHN, J. 1968a. Numbers of nitrifying bacteria in some Rhodesian soils under natural grass and improved pastures. *J. appl. Ecol.*, **5**, 291-300.

MEIKLEJOHN, J. 1968b. New nitrogen fixers from Rhodesian soils. *Trans. 9th int. Cong. Soil Sci.*, **2**, 141-149.

MEREDITH, R. M. 1965. A review of the responses to fertilizer of the crops of northern Nigeria. *Samaru misc. Pap. No.* 4.

MEYER, J. A. AND DUPRIEZ, G. L. 1959. The amounts of nitrogen and other nutrient elements supplied to the soil by rainwater in the Belgian Congo, and their agricultural significance. *Proc. 3rd inter-Afr. Soils Conf., Dalaba*, **1**, 495-499.

MEYER, J. A. AND PAMPFER, E. 1959. Nitrogen content of rainwater collected in the humid Central Congo basin. *Nature, London*, **184**, 717-718.

MILLS, W. R. 1953. Nitrate accumulation in Uganda soils. *E. Afr. agric. J.*, **19**, 53-54.

MILNE, M. K. 1968. *Annotated Bibliographies on Soil Fertility in West Africa.* Commonw. Bur. Soils, Harpenden, England.

MOLINA, J. A. AND ROVIRA, A. D. 1964. The influence of plant roots on autotrophic nitrifying bacteria. *Can. J. Microbiol.*, **10**, 249-257.

MONNIER, G. 1965. Action des matieres organiques sur la stabilite structural des sols (deuxieme partie). *Ann. agron.*, **16**, 471-534.

MONOD, T. AND TOUPET, C. 1961. Land use in the Sahara-Sahel region, pp. 239-253 in *A History of Land Use in the Arid Regions*, ed. Stamp. L. D. UNESCO, Paris.

MOORE, A. W. 1960. The use of nitrifiable nitrogen as an index of nitrogen availability in a tropical soil. *W. Afr. J. biol. Chem.*, **3**, 67-69.

MOORE, A. W. 1962. The influence of a legume on soil fertility under a grazed tropical pasture. *Emp. J. expl. Agric.*, **30**, 239-248.

MOORE, A. W. 1963a. Nitrogen fixation in latosolic soil under grass. *Pl. Soil.*, **10**, 127-138.
MOORE, A. W. 1963b. Occurrence of non-symbiotic nitrogen-fixing micro-organisms in Nigerian soils. *Pl. Soil*, **19**, 385-395.
MOORE, A. W. AND AYEKE, C. A. 1965. HF-extractable ammonium nitrogen in four Nigerian soils. *Soil Sci.*, **99**, 335-338.
MOREL, R. AND QUANTIN, P. 1964. (Fallowing and soil regeneration under the Sudan-Guinea climate in Central Africa.) *Agron. trop., Paris*, **19**, 105-136.
MOREL, R. AND QUANTIN, P. 1972. Observations sur l'evolution a long terme de la fertilite des sols cultives a Grimari (Republique Centrafricaine). Resultats d'essais de culture mecanisee semi-intensive, sur des sols rouges ferrallitiques moyennement desatures en climat soudano-guineen d'Afrique Centrale. *Agron. trop., Paris*, **27**, 667-739.
MORGAN, W. B. AND PUGH, J. C. 1969. *West Africa*. Methuen, London.
MORTIMORE, M. J. 1971. Population densities and systems of agricultural land-use in northern Nigeria. *Niger. geog. J.*, **14**, 3-15.
MOSS, R. P. 1968. Soils, slopes and surfaces in tropical Africa, pp. 29-60 in *The Soil Resources of Tropical Africa*, ed. Moss, R. P. Cambridge Univ. Press.
MOUREAUX, C. 1967. Influence de la temperature et de l'humidite sur les activities biologique de quelques sols Ouest-Africaine. *Cah. ORSTOM, Ser. pedol.*, **5**, 393-420.
MOUREAUX, C. AND FAUCK, R. 1967. Influence d'un exces d'humidite temporaire sur quelques sols de l'ouest Africaine. *Cah. ORSTOM, Ser. pedol.*, **5**, 103-113.
MUNRO, P. E. 1966. Inhibition of nitrite-oxidizers by roots of grass. Inhibition of nitrifiers by grass root extracts. *J. appl. Ecol.*, **3**, 227-230, 231-238.
MUNSON, R. D. AND STANFORD, G. 1955. Predicting nitrogen fertilizer needs of Iowa soils. IV. Evaluation of nitrogen production as a criterion of nitrogen availability. *Proc. Soil Sci. Soc. Am.*, **19**, 464-468.
NALOVIC, L. AND PINTA, M. 1972. Recherches sur les elements traces dans les sols tropicaux: etude de quelques sols du Cameroun. *Geoderma*, **7**, 249-267.
NETTING, R. McC. 1968. *Hill Farmers of Nigeria*. Univ. Washington Press.
NICOU, R. 1969. Action du labour sur la porosite. IRAT/Senegal, rapp. ann. division des Techn. cultivales (quoted by Charreau and Nicou, 1971c).
NICOU, R. AND POULAIN, J. F. 1967. La fumure minerale du niebe au Senegal. *Colloque sur la fertilite des sols tropicaux, Tananarive*, **1**, 720-730.
NICOU, R., SEGUY, L. AND HADDAD, G. 1970. Comparaison de l'enracinement de quatre varieties de riz pluvial en presence ou absence de travail du sol. *Agron. trop., Paris*, **25**, 639-659.
NICOU, R. AND THIROUIN, H. 1968. Mesures sur la porosite et l'enracinement. Premiers resultats. IRAT/Senegal mimeo.
NIQUEX, M. 1959. Choix de varietes d'arachides au Tchad. I. Regions a faible pluviometrie. *Agron. trop., Paris*, **14**, 490-502.
NOMMICK, H. 1965. Ammonium fixation and other reactions involving a nonenzymatic immobilization of mineral nitrogen in soil, pp. 198-258 in *Soil Nitrogen*, ed. Bartholomew, W. V. and Clark, F. E. Am. Soc. Agron., Madison, Wisconsin.
NORMAN, D. W. 1967. An economic study of three villages in Zaria Province. I. Land and labour relationships. *Samaru misc. Pap. No.* 19.
NORMAN, D. W. 1970. Intercropping. *Proc. Seminar on Intercropping. Ford Foundation IITA/IRAT, Ibadan*, 1970.
NORMAN, D. W. 1975. Dryland farming among the Hausa in the north of Nigeria, in *Tradition and Change in Agriculture*, ed. R. D. Stephens (in press).
NUTMAN, P. S. 1965. Symbiotic nitrogen fixation, pp. 360-383 in *Soil Nitrogen*, ed. Bartholomew, W. V. and Clark, F. E. Am. Soc. Agron., Madison, Wisconsin.
NYE, P. H. 1950. The relation between nitrogen responses, previous soil treatment and the C/N ratio in soils of Gold Coast savanna areas. *Trans. 4th int. Cong. Soil Sci.*, **1**, 246-249.
NYE, P. H. 1951. Studies on the fertility of Gold Coast soils. II. The nitrogen status of the soils. *Emp. J. expl. Agric.*, **19**, 275-282.
NYE, P. H. 1952a. Studies on the fertility of Gold Coast soils. III. The phosphate status of the soils. *Emp. J. expl. Agric.*, **20**, 47-55.

NYE, P. H. 1952b. Studies on the fertility of Gold Coast soils. IV. The potassium and calcium status of the soils and the effects of mulch and kraal manure. *Emp. J. expl. Agric.*, **20**, 227-233.

NYE, P. H. 1953. A survey of the value of fertilizers to the food-farming areas of the Gold Coast. II. The granitic soils of the far north. *Emp. J. expl. Agric.*, **21**, 262-274.

NYE, P. H. 1954. A survey of the value of fertilizers to the food farming areas of the Gold Coast. III. The Voltaian sandstone region and the southern maize areas. *Emp. J. expl. Agric.*, **22**, 42-54.

NYE, P. H. 1958a. The relative importance of fallows and soils in storing plant nutrients in Ghana. *J. W. Afr. Sci. Assoc.*, **4**, 31-49.

NYE, P. H. 1958b. Rep. Conf. Directors overseas Dept. Agric., Colonial Office Public. No. 531, pp. 43-47.

NYE, P. H. 1959. The level of humus under the system of shifting cultivation. *Proc. 3rd inter-Afr. Soils Conf., Dalaba*, **1**, 525-529.

NYE, P. H. 1963. Soil analysis and the assessment of fertility in tropical soils. *J. Sci. Food Agric.*, **14**, 277-280.

NYE, P. H. AND BERTHEUX, M. H. 1957. The distribution of phosphorus in forest and savanna soils of the Gold Coast and its agricultural significance. *J. agric. Sci., Camb.*, **49**, 141-159.

NYE, P. H. AND FOSTER, W. N. M. 1961. The relative uptake of phosphorus by crops and natural fallow from different parts of their root zone. *J. agric. Sci., Camb.*, **56**, 299-306.

NYE, P. H. AND GREENLAND, D. J. 1960. *The Soil under Shifting Cultivation*. Commonw. Bur. Soils Tech. Comm. No. 51, Harpenden, England.

NYE, P. H. AND STEPHENS, D. 1962. Soil Fertility, pp. 127-143 in *Agriculture and Land Use in Ghana*, ed. Wills, J. B. Oxford Univ. Press.

OBIHARA, C. H., BAWDEN, M. G. AND JUNGERIUS, P. D. 1964. The Anambra-Do rivers area. Soil Survey Memoir No. 1, Min. of Agric., Eastern Nigeria.

ODU, C. T. I. 1967. *Aerobic nitrogen transformations in some soils of the derived savanna.* Ph. D. thesis, Univ. of Ibadan.

ODU, C. T. I. AND ADEOYE, K. B. 1970. Heterotrophic nitrification in soils—a preliminary investigation. *Soil Biol. Biochem.*, **2**, 41-45.

ODU, C. T. I. AND VINE, H. 1968a. Non-symbiotic nitrogen fixation in a savanna soil of low organic matter content. *Proc. Symp. Isotopes & Radiation in Soil Organic Matter Studies, Vienna*, 1968, pp. 335-350.

ODU, C. T. I. AND VINE, H. 1968b. Transformation of ^{15}N-tagged ammonium and nitrate in some savanna soil samples. *Proc. Symp. Isotopes* & *Radiation in Soil Organic Matter Studies, Vienna*, 1968, pp. 351-361.

OFOMATA, G. E. K. 1964. Soil erosion in the Enugu region of Nigeria. *Afr. Soils*, **9**, 289-348.

OFOMATA, G. E. K. 1965. Factors of soil erosion in the Enugu area of Nigeria. *Niger. geogr. J.*, **8**, 45-59.

OFORI, C. S. 1973. The effect of ploughing and fertilizer application on the yield of cassava (*Manihot esculenta* Crantz). *Ghana J. agric. Sci.*, **6**, 21-24.

OFORI, C. S. AND NANDY, S. 1969. The effect of method of soil cultivation on yield and fertilizer response of maize grown on a forest ochrosol. *Ghana J. agric. Sci.*, **2**, 19-24.

OFORI, C. S. AND POTAKEY, V. A. 1965. Effect of fertilizers on yield of crop under mechanical cultivation in the coastal savanna area in Ghana. *Proc. OAU/STRC symposium on maintenance and improvement of soil fertility, Khartoum.*

OLIVER, R. 1972. Etude du statut phosphorique des sols de Madagascar (3e partie). Document IRAM, no. 313 (quoted by Pichot and Roche, 1972).

OLLAGNIER, M. AND GILLIER, P. 1965. (Comparison between foliar diagnosis and soil analysis in determining the fertilizer requirements of groundnuts in Senegal.) *Oleagineux*, **20**, 513-516.

OLLAGNIER, M. AND PREVOT, P. 1958a. The Chief mineral deficiencies of groundnuts, oil palm and coconut palm in French Africa. *Qual. Plant. Mater. veg.*, **3-4**, 550-568.

OLLAGNIER, M. AND PREVOT, P. 1958b. Experimentation sur le phosphate bicalcique et le phospal. *Oleagineux*, **13**, 589-594.

OLLAT, C. AND COMBEAU, A. 1960. Methods of determining the exchange capacity and pH of a soil. Relationship between the absorbent complex and pH. *Afr. Soils*, **5**, 373-380.

OMOTOSO, T. I. AND WILD, A. 1970. Content of inositol phosphates in some English and Nigerian soils. *J. Soil Sci.*, **21**, 216-223.

OTENG, J. W. AND ACQUAYE, D. K. 1971. Studies on the availability of phosphorus in representative soils of Ghana. I. Availability tests by conventional methods. *Ghana J. agric. Sci.*, **4**, 171-183.

PALMER, J. E. S. 1958. Soil conservation at Mokwa, Northern Nigeria. *Trop. Agric.*, Trin., **35**, 34-40.

PALMER, J. L. AND GOLDSWORTHY, P. R. 1971. Fertilizer trials with sprayed and unsprayed cotton in Nigeria. *Expl. Agric.*, **7**, 281-287.

PAPADAKIS, J. 1965. *Crop ecologic survey in West Africa*. FAO, Rome.

PATRICK, W. H. AND MAHAPATRA, J. C. 1968. Transformation and availability to rice of nitrogen and phosphorus. *Adv. Agron.*, **20**, 323-359.

PEREIRA, H. C., CHENERY, E. M. AND MILLS, W. R. 1954. The transient effect of grasses on the structure of tropical soils. *Emp. J. expl. Agric.*, **22**, 148-160.

PEYVE, Ya. V. 1963. Microelements in the savanna soils of the Republic of Mali. *Soviet Soil Sci.* (transl. of *Pochvovedeniye*) no. 11, 1050-1053.

PHILLIPS, J. 1959. *Agriculture and Ecology in Africa*. Faber and Faber, London.

PIAS, J. 1960. *Sols de la Region Est du Tchad, planis de piedmont, massifs du Ouaddai et de l'Enneda*, Vol. I. ORSTOM, Tchad.

PICHOT, J. AND ROCHE, P. 1972. Phosphore dans les sols tropicaux. *Agron. trop., Paris*, **27**, 939-965.

PICHOT, J., TRUONG, B. AND BURDIN, S. 1973. Evolution du phosphore dans un sol ferrallitique soumis a differents traitements agronomiques. *Agron. trop., Paris*, **28**, 131-146.

PIERI, C. 1967a. Bilan des recherches sur la fumure phosphate au Mali. *Colloque sur la fertilite des sols tropicaux, Tananarive*, **1**, 1139-1148.

PIERI, C. 1967b. Etude de l'erosion et du ruissellement a Sefa au cours de l'anee 1965. *Colloque sur la fertilite des sols tropicaux, Tananarive*, **2**, 1302-1315.

PIERI, C. 1971. Survey of the fertilization trials on rainfed cereals in Mali from 1954 to 1970. IRAT publication (translated STRC/OAU—J.P. 26, Dakar).

PIERI, C. 1973. La fumure des cereales de culture seche en Republique du Mali. Premier essai de synthese. *Agron. trop., Paris*, **28**, 751-766.

PINTA, M. AND OLLAT, C. 1961. Recherches physico-chimiques des elements-traces dans les sols tropicaux. I. Etude de quelques sols du Dahomey. *Geochim. Cosmochim. Acta*, **25**, 14-23.

PORTERES, R. 1952. The agricultural crisis in Senegal. *Afr. Soils*, **3**, 40-51.

POULAIN, J. F. 1965. Contribution a l'etude des mechanisms d'action de la fumure verte—effets sur le sol et les rendements des cultures. *Proc. 2nd meeting on soil fertility and fertilizer use in W. Africa*, **1**, 79-98.

POULAIN, J. F. 1967. Resultats obtenus avec les engrais et les amendements calciques. Acidification des sols et correction. *Colloque sur la fertilite des sols tropicaux, Tananarive*, **1**, 469-489.

POULAIN, F. J. AND ARRIVETS, J. 1972. Effet des principaux elements fertilisants autre que l'azote sur les rendements de cultures vivrieres de base: sorgho-mil-mais du Senegal et en Haute-Volta. *Afr. Soils*, **17**, 189-214.

POULAIN, J. F. AND MARA, M. 1965. Comparaison de l'action de differents engrais phosphates utilisables au Senegal. *Proc. 2nd meeting on soil fertility and fertilizer use in W. Africa*, **2**, 248-266.

POULAIN, J. F. AND TOURTE, R. 1970. Effects of deep preparation of dry soil on yields from millet and sorghum to which nitrogen fertilizers have been added (sandy soil from a dry tropical area). *Afr. Soils*, **15**, 553-586.

PREVOT, P. AND MARTIN, G. 1964. Comparaison des fumures organique et minerales pour l'arachide en Haute-Volta. *Oleagineux*, **19**, 533-537.

PREVOT, P. AND OLLAGNIER, M. 1959. Epuisement du sol et effet des fumures dans un assolement continu arachide-mil. *Oleagineux*, **14**, 423-431.

PULLAN, R. A. 1961. Report on the reconnaissance and detailed soil surveys of the Birnin Kebbi area. Samaru Soil Survey Bull. No. 17. Inst. Agric. Res., Samaru, Nigeria.

PULLAN, R. A. 1962a. A report on the reconnaissance soil survey of the Nguru-Hadejia-Gumel area with special reference to the establishment of an experimental farm. Samaru Soil Survey Bull. No. 18. Inst. Agric. Res., Samaru, Nigeria.

PULLAN, R. A. 1962b. A report on the reconnaissance soil survey of the Azare (Bauchi) area with special reference to the establishment of an experimental farm. Samaru Soil Survey Bull. No. 19. Inst. Agric. Res., Samaru, Nigeria.

PULLAN, R. A. 1969. The soil resources of West Africa, pp. 147-191 in *Environment and Land Use in Africa,* ed. Thomas, M. F. and Whittington, G. W. Methuen, London.

PULLAN, R. A. 1970. The soils, soil landscapes and geomorphological evolution of a metasedimentary area in Northern Nigeria. *Dept. Geogr., Univ. Liverpool, Res. Pap. No.* 6.

PURSEGLOVE, J. W. 1968. *Tropical Crops: Dicotyledons.* Longmans, London.

PURSEGLOVE, J. W. 1972. *Tropical Crops: Monocotyledons.* Longmans, London.

QUANTIN, P. 1965. *Les sols de la Republique Centrafricaine.* ORSTOM publication.

RADWANSKI, S. A. AND WICKENS, G. E. 1967. The ecology of *Acacia albida* on mantle soil in Zalingei, Jebel Marra, Sudan. *J. appl. Ecol.,* **4,** 569-579.

RAINS, A. B. 1963. Grassland research in Northern Nigeria, 1952-62. *Samaru misc. Pap. No.* 1.

RAMSAY, J. M. AND INNES, R. R. 1963. Some quantitative observations on the effects of fire on the Guinea Savanna vegetation of northern Ghana over a period of eleven years. *Afr. Soils,* **8,** 41-86.

RANKAMA, K. AND SAHAMA, T. G. 1950. *Geochemistry.* Chicago Univ. Press.

RICE, E. L. 1964. Inhibition of nitrogen-fixing and nitrifying bacteria by seed plants, I. *Ecology,* **45,** 824-837.

RICE, E. L. 1965. Inhibition of nitrogen fixing and nitrifying bacteria by seed plants, II. *Physiol. Planta,* **18,** 255-268.

RICHARD, L. 1967. Evolution de la fertilite en culture cotonniere intensive. *Colloque sur la fertilite des sols tropicaux, Tananarive,* **2,** 1437-1471.

RICHARD, L. 1970. Role de la fertilisation de la culture cotonniere dans l'intensification de la production agricole. *Proc. 9th Cong. int. Potash Inst., Antibes,* pp. 241-248.

RICHARDSON, H. L. 1938. The nitrogen cycle in grassland soils: with especial reference to the Rothamsted Park grass experiment. J. *agric. Sci., Camb.,* **28,** 72-121.

RILEY, D. AND YOUNG, A. 1966. *World Vegetation.* Cambridge Univ. Press.

ROBINSON, J. B. 1963. Nitrification in a New Zealand grassland soil. *Pl. Soil,* **19,** 173-183.

ROBINSON, J. B. D. 1957. The critical relationship between soil moisture content in the region of the wilting point and the mineralization of natural soil nitrogen. *J. agric. Sci., Camb.,* **49,** 100-105.

ROBINSON, J. B. D. 1968. Measuring soil nitrogen available to crops in East Africa. *E. Afr. agric. For. J.,* **33,** 269-280.

ROCHE, P. 1970. Soil fertility problems. *Agron. trop., Paris,* **25,** 875-892.

ROOSE, E. 1967. Dix annees de mesure de l'erosion et du ruissellement au Senegal. *Agron. trop., Paris,* **22,** 123-152.

ROOSE, E. 1973. *Dix-sept annees de mesures experimentales de l'erosion et du ruissellement sur un sol ferrallitique sableux de basse Cote d'Ivoire.* Thesis pour Docteur-Ingenieur, Universite d'Abidjan; pub. ORSTOM, Abidjan.

ROOSE, E. AND BERTRAND, R. 1971. Contribution a l'etude de la methode des bandes d'arret pour lutter contre l'erosion hydrique en Afrique de l'Ouest. Resultats experimentaux et observations sur le terrain. *Agron. trop., Paris,* **26,** 1270-1283.

ROUX, J. B. AND GUTKNECHT, J. 1956. Essais de fumure du cotonnier au Tchad. *Cot. Fib. trop.,* **11,** 137-152.

ROUNCE, N. V. AND MILNE, G. 1936. The unsuitability of certain virgin soils to the growth of grain crops. *E. Afr. agric. J.,* **2,** 145-148.

RUSSELL, E. W. 1968a. The place of fertilizers in food crop economy of tropical Africa. *Proc. Fertil. Soc. Lond., No.* 101.

RUSSELL, E. W. 1968b. Some agricultural problems of semi-arid areas, pp. 121-135 in *The Soil Resources of Tropical Africa,* ed. Moss, R. P. Cambridge Univ. Press.

RUSSELL, E. W. 1971. Soil structure; its maintenance and improvement. *J. Soil Sci.,* **22,** 137-151.

Russell, E. W. 1973. *Soil Conditions and Plant Growth* (10th edition). Longmans, London.
Ruthenberg, H. 1971. *Farming Systems in the Tropics*. Clarendon Press, Oxford.
Sauchelli, V. 1969. *Trace Elements in Agriculture*. Van Nostrand Reinhold Co., New York.
Saunder, D. H. and Grant, P. M. 1962. Rate of mineralization of organic matter in cultivated Rhodesian soils. *Trans. int. Soil Sci. Soc. Comm. IV & V, New Zealand*, 235-239.
Scarsbrook, C. E. 1965. Nitrogen availability, pp. 481-502 in *Soil Nitrogen*, ed. Bartholomew, W. V. and Clark, F. E. Am. Soc. Agron., Madison, Wisconsin.
Schilling, R. 1965. Effet residual et cumulatif des engrais au Senegal. *Proc. 2nd meeting on soil fertility and fertilizer use in W. Africa*, **2**, 273-285.
Schlippe, P. de 1956. *Shifting Cultivation in Africa: the Zande System of Agriculture*. Routledge and Kegan Paul, London.
Schoch, P. G. 1966. Influence sur l'evaporation potentielle d'une strate arboree au Senegal et consequences agronomiques. *Agron. trop., Paris*, **21**, 1283-1290.
Schutte, K. 1954. A survey of plant minor element deficiencies in Africa. *Afr. Soils*, **3**, 285-297.
Scott, R. M. 1962. Exchangeable bases of mature well-drained soils in relation to rainfall in East Africa. *J. Soil Sci.*, **13**, 1-9.
Semb, G. and Robinson, J. B. D. 1969. The natural nitrogen flush in different arable soils and climates in East Africa. *E. Afr. agric. For. J.*, **34**, 350-370.
Siband, P. 1972. Etude de l'evolution des sols sous culture traditionelle en Haute-Casamance. Principaux resultats. *Agron. trop., Paris*, **27**, 574-591.
Simpson, J. R. 1960. The mechanism of surface nitrate accumulation on a bare fallow soil in Uganda. *J. Soil Sci.*, **11**, 45-60.
Simpson, J. R. 1961. The effects of several agricultural treatments on the nitrogen status of a red earth in Uganda. *E. Afr. agric. For. J.*, **26**, 158-163.
Singh, K. 1961. Value of bush, grass and legume fallow in Ghana. *J. Sci. Fd. Agric.*, **12**, 160-168.
Smyth, A. J. and Montgomery, R. F. 1962. *The Soils and Land Use of Central Western Nigeria*. The Government Printer, Ibadan, Western Nigeria.
Stamp, C. D. 1953. *Africa: A Study in Tropical Development*. Wiley, New York.
Steele, W. M. 1972. *Cowpeas in Nigeria*. Ph. D. Thesis, Reading Univ.
Stephen, I. 1953. A petrographic study of a tropical black earth and grey earth from the Gold Coast. *J. Soil Sci.*, **4**, 211-219.
Stephens, D. 1959. Field experiments with trace elements on annual food crops in Ghana. *Emp. J. expl. Agric.*, **27**, 324-332.
Stephens, D. 1960a. Fertilizer trials on peasant farms in Ghana. *Emp. J. expl. Agric.*, **28**, 1-15.
Stephens, D. 1960b. Fertilizer experiments with phosphorus, nitrogen and sulphur in Ghana. *Emp. J. expl. Agric.*, **28**, 151-164.
Stephens, D. 1960c. Three rotation experiments with grass fallows and fertilizers. *Emp. J. expl. Agric.*, **28**, 165-178.
Stephens, D. 1962. Upward movement of nitrate in a bare soil in Uganda. *J. Soil Sci.*, **13**, 52-59.
Stephens, D. 1967. Effects of grass fallow treatments in restoring fertility of Buganda clay loam in South Uganda. *J. agric. Sci., Camb.*, **68**, 391-403.
Stevenson, F. J. 1965. Origin and distribution of nitrogen in soil, pp. 1-42 in *Soil Nitrogen*, ed. Bartholomew, W. V. and Clark, F. E. Am. Soc. Agron., Madison, Wisconsin.
Stiven, G. 1952. Production of antibiotic substances by the roots of grasses (*Trachypogon phimosus* H.B.X. Nees) and of *Pentanesia variabilis* (E. May) Harv. (*Rubiaceae*). *Nature, London*, **170**, 712-713.
Stockinger, K. R. 1970. Environmental factors limiting yield of the major cereals in northern Nigeria. *Afr. Soils*, **15**, 481-484.
Sys, C. 1969. The soils of Central Africa in the American classification—7th Approximation. *Afr. Soils*, **14**, 5-44.
Theron, J. J. 1951. The influence of plants on the mineralization of nitrogen and the maintenance of organic matter in soil. *J. agric. Sci., Camb.*, **41**, 289-296.

THEVIN, C. 1967. Exemple d'une etude regionale de fertilisation en culture cotonniere (Kandi-Banikoara-Dahomey). *Colloque sur la fertilite des sols tropicaux, Tananarive,* **1**, 211-220.

THOMANN, C. 1964. Les differentes fractions humiques de quelques sols tropicaux de l'ouest Africain. *Cah. ORSTOM, Ser. pedol.,* **2**, 2, 43-79.

THOMAS, M. F. 1969. Geomorphology and land classification in tropical Africa, pp. 103-145 in *Environment and Land Use in Africa,* ed. Thomas, M. F. and Whittington, G. W. Methuen, London.

THOMPSON, B. 1965. *The Climate of Africa.* Oxford Univ. Press.

THORNTON, I. 1965. Nutrient content of rainwater in the Gambia. *Nature, London,* **205**, 1025.

TOMLINSON, P. R. 1965. Soils of Northern Nigeria. *Samaru misc. Pap. No.* 11.

TOURTE, R. 1971. Themes legers. Themes lourds. Systemes intensifs. Voies differentes ouvertes au developpement agricole du Senegal. *Agron. trop., Paris,* **26**, 632-671.

TOURTE, R., NICOU, R. AND BOULIEU, A. 1963. Quelques techniques de culture des mils et sorghos au Senegal. Possibilities de la culture mecanique. *Agron. trop., Paris,* **18**, 65-72.

TOURTE, R., NICOU, R., CHARREAU, C., POCTHIER, G. AND POULAIN, J. F. 1967. La fumure minerale "etalee" au Senegal. Comparaison avec la fumure annuelle. *Colloque sur la fertilite des sols tropicaux, Tananarive,* **1**, 1058-1075.

TOURTE, R., VIDAL, P., JACQUINOT, L., FAUCHE, J. AND NICOU, R. 1964. Bilan d'une rotation quadriennale sur sole de regeneration au Senegal. *Agron. trop., Paris,* **19**, 1033-1072.

TURNER, D. J. 1966. An investigation into the causes of low yield in late-planted maize. *E. Afr. agric. For. J.,* **31**, 249-260.

TRAPNELL, C. G. 1959. Ecological results of woodland burning experiments in Northern Rhodesia. *J. Ecol.,* **47**, 129-168.

UDO, E. J. AND UZU, F. O. 1972. Characteristics of phosphorus adsorption by some Nigerian soils. *Proc. Soil Sci. Soc. Am.,* **36**, 879-883.

USDA, 1960. *Soil Classification, A Comprehensive System, 7th Approximation.* U.S. Dept. Agric., Washington.

USAID, 1962. Report on soil and agricultural survey of Sene-Obosum river basins, Ghana.

VAILLE, J. 1970. Fertilization of sorghum in northern Cameroun. *Afr. Soils,* **15**, 77-83.

VAN RAAY, J. G. T. AND LEEUW, P. N. DE 1970. The importance of crop residues as fodder: a resource analysis in Katsina Province, Nigeria. *Tijdschr. econ. soc. Geogr.,* **61**, 137-147.

VEIHMEYER, F. J. AND HENDRIKSON, A. H. 1948. Soil density and root penetration. *Soil Sci.,* **65**, 487-493.

VELLY, J. 1972. Fertilisation potassique des sols tropicaux. *Agron. trop., Paris,* **27**, 966-977.

VERNEY, R. AND WILLIAME, P. 1965. Results of studies of erosion on experimental plots in Dahomey. *Proc. OAU/STRC symposium on the maintenance and improvement of soil fertility, Khartoum.*

VIDAL, P. AND FAUCHE, J. 1961. Annals Cent. Rech. agron., Bambey, No. 20, 104-116.

VIDAL, P. AND FAUCHE, J. 1962. Quelques aspects de la dynamique des elements mineraux d'un sol dior soumis a differentes jacheres. Premiers resultats. *Agron. trop., Paris.,* **17**, 828-840.

VINCENT, J. M. 1965. Environmental factors in the fixation of nitrogen by the legume, pp. 384-435 in *Soil Nitrogen,* ed. Bartholomew, W. V. and Clark, F. E. Am. Soc. Agron., Madison, Wisconsin.

VINE, H. 1953. Experiments on the maintenance of soil fertility at Ibadan, Nigeria, 1922-51. Pt. I. Introduction and review of the period 1922-33. *Emp. J. expl. Agric.,* **21**, 65-85.

VINE, H. 1968. Developments in the study of soils and shifting agriculture in tropical Africa, pp. 89-119 in *The Soil Resources of Tropical Africa,* ed. Moss, R. P. Cambridge Univ. Press.

VISSER, S. A. 1966. Annual variation in the distribution and activity of different groups of microorganisms in a Nigerian groundnut soil. *W. Afr. J. biol. appl. Chem.,* **9**, 20-24.

WALKER, H. O. 1962. Weather and Climate, pp. 7-50 in *Agriculture and Land Use in Ghana*, ed. Wills, J. B. Oxford Univ. Press.
WALTER, M. W. 1967. Length of the rainy season in Nigeria. *Niger. geog. J.*, **10**, 123.
WATSON, K. A. 1964. Fertilizers in Northern Nigeria. Current utilization and recommendations for their use. *Afr. Soils*, **9**, 5-20.
WATSON, K. A. AND GOLDSWORTHY, P. R. 1964. Soil fertility investigations in the middle belt of Nigeria. *Emp. J. expl. Agric.*, **32**, 290-302.
WEBB, R. A. 1954. The mineral deficiences of some Gambian soils. *Proc. 2nd inter-Afr. Soils Conf.*, **1**, 419-424.
WEINHARD, P., BALANDREAU, J., RINAUDO, G. ET AL. 1971. Non-symbiotic nitrogen fixation in the rhizosphere of some non-leguminous tropical plants. *Rev. Ecol. Biol. Sol*, **8**, 367-373.
WERTS, R. AND BOUYER, S. 1965. Research carried out by IRAT on soil fertilization in Dahomey. *Proc. OAU/STRC symposium on the maintenance and improvement of soil fertility, Khartoum.*
WERTS, R. AND BOUYER, S. 1967. Etudes recentes sur la fertilisation des sols realisees par l'IRAT au Dahomey. *Colloque sur la fertilite des sols tropicaux, Tananarive*, **1**, 601-612.
WEST, O. 1965. Fire in vegetation and its use in pasture management. C.A.B. mimeo publication.
WETSELAAR, R. 1961. Nitrate distribution in tropical soils. II. Extent of capillary accumulation of nitrate during long dry period. *Pl. Soil*, **15**, 121-133.
WETSELAAR, R. AND HUTTON, J. T. 1963. The ionic composition of rainwater at Katherine, N. T. and its part in the cycling of plant nutrients. *Aust. J. agric. Res.*, **14**, 319-329.
WETSELAAR, R. AND NORMAN, M. J. T. 1960. Recovery of available soil nitrogen by annual fodder crops at Katherine, N. T. *Aust. J. agric. Res.*, **11**, 693-704.
WHITE, S. 1941. Agricultural economy of the hill pagans of Dikwa Emirate, Cameroons. *Emp. J. expl. Agric.*, **9**, 65-72.
WHITEHEAD, D. C. 1964. Soil and plant-nutrition aspects of the sulphur cycle. *Soils Fertil.*, **27**, 1-8.
WILD, A. 1958. The phosphate content of Australian soils. *Aust. J. agric. Res.*, **9**, 193-204.
WILD, A. 1961. A pedological study of phosphorus in 12 soils derived from granite. *Aust. J. agric. Res.*, **12**, 286-299.
WILD, A. 1971. The potassium status of soils in the savanna zone of Nigeria. *Expl. Agric.*, **7**, 257-70.
WILD, A. 1972a. Mineralization of soil nitrogen at a savanna site in Nigeria. *Expl. Agric.*, **8**, 91-97.
WILD, A. 1972b. Nitrate leaching under bare fallow at a site in northern Nigeria. *J. Soil. Sci.*, **23**, 315-324.
WILKINSON, G. E. 1970. The infiltration of water into Samaru soils. *Samaru agric. Newsletter*, **12**, 81-83. Inst. Agric. Res., Samaru, Nigeria. (Also *Trop. Agric., Trin.*, **52**, 97-103. 1975)
WILLIAMS, C. H. AND LIPSETT, J. 1961. Fertility changes in soils cultivated for wheat in southern New South Wales. *Aust. J. agric. Res.*, **12**, 612-629.
WISCHMEIER, W. H. AND SMITH, D. D. 1960. A universal soil-loss estimating equation to guide conservation farm planning. *Trans. 7th int. Cong. Soil Sci.*, **1**, 418-425.
WOODRUFF, J. R. AND KAMPRATH, E. J. 1965. Phosphorus adsorption maximum as measured by the Langmuir isotherm and its relationship to phosphorus availability. *Proc. Soil. Sci. Soc. Am.*, **29**, 148-150.
YOUNGE, O. R. AND PLUCKNETT, D. L. 1966. Quenching the high phosphorus fixation of Hawaiian Latosols. *Proc. Soil Sci. Soc. Am.*, **30**, 653-655.

SUBJECT INDEX